DAMS AND DEVELOPMENT IN **CHINA**

CONTEMPORARY ASIA IN THE WORLD

CONTEMPORARY ASIA IN THE WORLD

DAVID C. KANG AND VICTOR D. CHA, EDITORS

This series aims to address a gap in the public-policy and scholarly discussion of Asia. It seeks to promote books and studies that are on the cutting edge of their respective disciplines or in the promotion of multidisciplinary or interdisciplinary research but that are also accessible to a wider readership. The editors seek to showcase the best scholarly and public-policy arguments on Asia from any field, including politics, history, economics, and cultural studies.

BRYAN TILT

DAMS AND DEVELOPMENT IN CHINA

The Moral Economy of Water and Power

COLUMBIA UNIVERSITY PRESS · NEW YORK

Columbia University Press
Publishers Since 1893
New York   Chichester, West Sussex
cup.columbia.edu

Library of Congress Cataloging-in-Publication Data

Tilt, Bryan, 1974–
    Dams and development in China : the moral economy of water and power / Bryan Tilt.
        pages cm. — (Contemporary Asia in the world)
    Includes bibliographical references and index.
    ISBN 978-0-231-17010-9 (cloth : alk. paper) — ISBN 978-0-231-17011-6
(pbk. : alk. paper) — ISBN 978-0-231-53826-8 (e-book)
    1. Dams—Social aspects—China—Yunan Xian. 2. Watershed management—
China. 3. Hydroelectric power plants—China. 4. Energy policy—China.
5. Economic development—Social aspects—China. I. Title.
    TC558.C5T55 2015
    333.91'40951—dc23

                                                                        2014017471

Columbia University Press books are printed on permanent and
durable acid-free paper.

This book is printed on paper with recycled content.
Printed in the United States of America

c 10 9 8 7 6 5 4 3 2 1
p 10 9 8 7 6 5 4 3 2

Cover image: © *Keren Su/Corbis*
Cover and book design: Lisa Hamm

References to websites (URLs) were accurate at the time of writing.
Neither the author nor Columbia University Press is responsible
for URLs that may have expired or changed since the manuscript
was prepared.

The antagonists in such contests . . . know each other's repertoire of practical action and discursive moves. There is, in other words, a kind of larger social contract that gives some order and limits to the conflict. . . . The limits and constraints characterizing conflict are never cut-and-dried to the participants. The antagonists are, all of them, continually prospecting new terrain—trying out new stratagems and wrinkles that threaten to change, and often do change, the shape of the "game" itself.

—James C. Scott, *The Art of Not Being Governed*

Displacement is possible only because the hydropower development companies aren't forced to look at the results of their actions. The hydropower development companies shouldn't be allowed to keep their distance. They should have to be shut in a room with the villagers that they have displaced. They should have to look those villagers in the eye.

—Dr. Li, Beijing-based sociologist

CONTENTS

ILLUSTRATIONS

FIGURES

TABLES

PREFACE

THIS BOOK examines the complex array of water-management decisions faced by Chinese leaders and the consequences of those decisions for local communities in the southwestern province of Yunnan. It focuses in particular on the construction of large dams to provide hydroelectricity to fuel the national economy, a process that provides alternative energy to further propel China's economic boom but also threatens sensitive ecological areas and displaces thousands of people who belong to economically and culturally marginalized groups. It is a story that involves natural-resource professionals, scientists and engineers, policy makers, domestic and international conservation organizations, and rural villagers.

The concept of the "moral economy," which figures prominently in this book, can be freighted with diverse and sometimes unintended meanings, and I wish to limit the potential for misunderstanding at the outset. I use the moral economy concept as an analytical framework in this book not to advance a particular agenda or to advocate for certain policy outcomes but rather to elucidate the normative choices that must be made when various important objectives—economic development, energy production, biodiversity conservation, and the protection of the rights of vulnerable people, among others—come into conflict. My aim is to show how different constituent groups use varying strategies, grounded in cultural and historical precedents, to advance quite different moral visions about how water resources should be managed.

China is home to approximately 25,000 large dams, a figure roughly equal to the number of dams in all other countries of the world combined.

The fact that China had become the global leader in dam construction had not escaped my attention as I worked on other topics, including pollution control and agricultural development, throughout the southwest region. In fact, during a field visit to southern Sichuan Province in 2006 I discovered that much of the out-migration in the township where I had done research for years was driven by the construction of the Jin'anqiao Hydropower Station on the Jinsha River, as the upper portion of the Yangtze is called, which for a time absorbed huge amounts of unskilled labor and allowed many villagers to send remittances back home to their families.

It was around that time that I discovered shared interests with scientists, both American and Chinese, from many other disciplines—including engineering, economics, geography, and hydrology—about dams and their effects on ecosystems and communities. In our conversations, we discovered that each of our disciplines had a basic understanding of how dams could affect ecosystems and communities—by changing the geomorphology and water quality of a river or by forcing farmers off their land and into towns and cities—but that we lacked a holistic understanding of the full range of impacts and their interrelationships. Since that time, I have taken enormous pleasure in learning about the theories, methods, and models of other disciplines and their application to water resources; I have also become convinced that the long-term solutions to such problems lie in these sorts of transnational and cross-disciplinary collaborative arrangements. This book reflects that sort of interdisciplinary engagement, an endeavor that I have approached less as a scholar than as a student, learning about river geomorphology, conservation biology and ecology, development economics, and domestic and international water governance.

It can be somewhat frustrating to work on a book-length study of energy and water-resources development in China because the policies and politics, fraught with a great deal of controversy, change almost daily. I'm not sure whether to feel appreciation or anger when I receive an email from a colleague or student with links to news articles about some new policy pronouncement or development plan; this change could mean something as simple as adding a note to my files or something as complicated and maddening as restructuring an entire chapter. In particular, developments related to the Nu River dams, which I discuss in detail in this book, have the back-and-forth quality of a tennis match: top

Chinese leaders have slowed or stopped the projects half a dozen times in the past ten years only to move them forward again. It goes without saying, therefore, that this book is not meant to be a definitive treatment of water-resource issues in China but rather a close look at a small corner of a vast and changing landscape.

ACKNOWLEDGMENTS

I owe a great intellectual debt to the people with whom I have collaborated over the years. Since 2006, I have worked with an interdisciplinary group of scientists on the Integrative Dam Assessment Model, with financial support from the Human and Social Dynamics Program at the U.S. National Science Foundation (Grants 0826752 and 0623087). Principal investigators and key collaborators on this project include Desiree Tullos, Aaron T. Wolf, Darrin Magee, Philip H. Brown, He Daming, Feng Yan, and Shen Suping. My work on the project, which has been both challenging and rewarding, has fundamentally enriched the way I see the issues that I present in this book. Some of the survey data that appear in places in this book constitute collective intellectual property, gathered through the hard work of many researchers and students, and I thank my colleagues for allowing me to use them here. The analysis and interpretation of these data, along with the conclusions and recommendations drawn from them, are mine alone and do not represent the consensus of the group.

I would also like to acknowledge the colleagues and students who either reviewed portions of this manuscript or provided substantial ideas and intellectual feedback along the way: Kelly Alley, Marco Clark, Ding Guiru, Du Fachun, Eric Foster-Moore, Brendan Galipeau, Francis Gassert, Edward Grumbine, Stevan Harrell, Hu Tao, Barbara Rose Johnston, Kelly Kibler, Elina Lin, Li Xiaoyue, Ralph Litzinger, Riall Nolan, Carlos Rojas, Sai Han, Edwin Schmitt, Janet Upton, Wang Hua, Yang Donghui, Zhang Haiyang, and Zhang Yong. I am grateful to Anne Routon at Columbia University Press for her unflagging support of this project. The maps that appear in the book and that help to contextualize the study sites for the reader were created by Brittany Albertson and Loretta Wardrip.

Many institutions provided material, financial, or logistical support during the course of research on this book, most notably the U.S.

National Science Foundation's Human and Social Dynamics Program, as noted earlier. I was also fortunate to conduct portions of the research for this book as a Fulbright Senior Research Scholar in China during 2012, which helped open many new doors and avenues of inquiry. At Pennsylvania State University, I participated in a workshop titled "China in Motion," organized by David Atwill and Kate Merkel-Hess and funded by the newly founded Confucius Institute at Penn State. The participants at that workshop, mostly historians, provided critical feedback and forgave my cursory understanding of historical currents that run wide and deep in southwestern China and Southeast Asia. I wish to express thanks to my colleagues in the School of Ethnology and Sociology at Minzu University of China, who provided an institutional home for me in Beijing. Special thanks are due to Professor Liu Mingxin, who graciously agreed to be my host and who introduced me to many colleagues who share my interest in shaping economic development in ways that recognize the value of human diversity and dignity.

Many other institutions provided support, including the Center for the Humanities at Oregon State University; the School of Language, Culture, and Society at Oregon State University; the East Asian Studies Program at Lewis and Clark College; the Asian International Rivers Center at Yunnan University; Yunnan Normal University; the Minority Nationalities Research Center at Minzu University; the Institute of Ethnology and Anthropology within the Chinese Academy of Social Sciences; the Center for Indigenous Peoples and Development at Yi-Shou University, Taiwan; the Woodrow Wilson International Center for Scholars; and the China Studies Program at the University of Washington.

I also thank the many people who participated in this research, from the professional resource managers and scientists in high-rise office complexes in Beijing to the villagers in Yunnan Province. Interacting with people and learning about their individual stories and their ambitions for the future are the most rewarding parts of conducting social science research. Because of the political sensitivity of portions of this research, I have chosen to protect the participants' anonymity by either using pseudonyms or referring to them only by their surnames. Half of the author's proceeds from this book will be donated to community-based development organizations in Southwest China.

As always, my family provided a great deal of support throughout this project; it was a rare pleasure to involve my wife, Jenna, and our two children in fieldwork and residency in China for six months in 2012 and to share the discovery process with them. I learned that my children—who enrolled in a Chinese school, tried strange food on a daily basis, and made new friends—are far more adaptable than I often given them credit for.

A WORD ON TERMINOLOGY

For Chinese terms, I use the pinyin system that is standard in the People's Republic of China, except for a few proper nouns that are commonly known by other spellings. This book focuses a great deal on two transboundary rivers that have their headwaters in China and flow through other riparian nations in Southeast Asia. When using non-Chinese foreign names, I have attempted to use the most widely accepted versions (e.g., Salween River instead of Thalwin River). In cases where confusion may arise, such as the use of place-names with multiple linguistic origins, I have tried to provide clarity in the endnotes. I have also included a list of terms in both pinyin and Chinese characters and a list of abbreviations that appear frequently in the text. Most of the interviews and surveys were conducted in Chinese. Translations of quotations from interviews, surveys, and publications in Chinese are mine unless otherwise noted.

ABBREVIATIONS

CCP	Chinese Communist Party
CDM	Clean Development Mechanism
CPIC	China Power Investment Corporation
EIA	environmental impact assessment
GDP	gross domestic product
GWh	gigawatt hour
HDI	Human Development Index
ICOLD	International Commission on Large Dams
IDAM	Integrative Dam Assessment Model
IRBM	integrated river-basin management
MEP	Ministry of Environmental Protection
MRC	Mekong River Commission
MWR	Ministry of Water Resources
NDRC	National Development and Reform Commission
NEPA	U.S. National Environmental Policy Act
NGO	nongovernmental organization
PRC	People's Republic of China
SIA	social impact assessment
TNC	The Nature Conservancy
UN	United Nations
UNDP	United Nations Development Program
UNEP	United Nations Environment Program
UNESCO	United Nations Educational, Scientific, and Cultural Organization
WCD	World Commission on Dams

DAMS AND
DEVELOPMENT
IN CHINA

1

THE MORAL ECONOMY
OF WATER AND POWER

AS ONE flies over northwest Yunnan Province in an airplane, skirting along the eastern edge of the Himalaya, there are points at which, depending on cloud cover, one can see all three of China's great Parallel Rivers (Sanjiang Bingliu) in a single glance: the Nu (known in Southeast Asia as the Salween), which cuts a path directly south into Myanmar (Burma); the Lancang (upper Mekong), which meanders through western Yunnan before passing through five other riparian nations in Southeast Asia; and the Jinsha, the headwaters of the Yangtze, the longest river in China. The view from the air is of a rugged landscape of glaciated mountain peaks and valleys crisscrossed by rivers. But it is only from the ground that one gains a sense of the region's tremendous biological and cultural heritage, the threats currently facing this heritage, and the multisided struggle to determine these rivers' future.

The Three Parallel Rivers region, a small corner of southwestern China, is home to 6,000 plant species and 80 species of rare or endangered animals, including treasured species such as the Yunnan snub-faced monkey (*Rhinopithecus bieti*), an extraordinary and infrequently encountered mammal that has become one charismatic symbol of the struggle to conserve what remains of this repository of biological diversity. It is also home to twenty-two of China's officially recognized minority nationalities (*minzu*). The people who live here, supporting themselves mostly by subsistence and small-scale market farming, are among the poorest in the nation. The region has become a focal point in the conflict between those who wish to develop China's rivers for their hydropower potential, including government agencies and hydropower corporations, and those who place a premium on preserving species richness and protecting the rights of vulnerable people.

The World Heritage Monitoring Center, which is part of the United Nations Environment Program (UNEP), has called the Three Parallel Rivers region an "epicenter of Chinese endemic species" (UNESCO 2003:4). This extensive area of varied microbiomes supported by a series of deep gorges and glaciated peaks, called the Hengduan Mountain Range, began rising skyward with the collision of the Indian tectonic plate and the Eurasian plate more than 50 million years ago. The mountains, like those of the main Himalayan range farther west, are still growing. Within a comparatively small land area, there are glaciers and scree, alpine meadow, alpine conifer, deciduous forest, cloud forest, mixed forest, savannah, and riparian habitats (Xu and Wilkes 2004). Moist air masses pushing in from the Indian Ocean, particularly during the summer monsoon season, deposit rain in the westernmost valleys before slowly petering out as they move inland, creating markedly different ecological conditions from one valley to the next. An overland trek of 50 kilometers from west to east will take a person from a lush biome with ferns and orchids to dry slopes covered in scrub and cacti.

The struggle to balance economic development and environmental protection increasingly involves both domestic and international players. In 2003, the United Nations Educational, Scientific, and Cultural Organization (UNESCO) designated fifteen protected areas in eight clusters, totaling nearly 1.7 million hectares, as World Heritage Sites (UNEP 2009). The designation includes approximately one million hectares of "core" protected areas and nearly 700,000 hectares of "buffer" areas in which limited human activity is allowed. The Nature Conservancy, working in close association with the Yunnan provincial government, is also active in land conservation and has advocated for turning Xianggelila County (Zhongdian changed its name to the mythical "Shangri-La" in 2003 in pursuit of tourism revenue) into a national park. Pudacuo National Park, a majestic landscape of mountains and alpine lakes located a few short kilometers away from the cobblestone lanes of Shangri-La Old Town, was established in 2007 to preserve the region as a "biodiversity hot spot," one of the richest reservoirs of flora and fauna on earth; it is the first in China to meet the standards of the International Union for the Conservation of Nature.

Around the world, most sensitive areas targeted for conservation represent a balancing act between the use of natural resources for human development and the imperative for environmental protection. In China, which over the past several decades has undergone economic and infrastructural development on an unprecedented scale and at a breakneck pace, the balancing act is particularly precarious. China's rivers hold massive undeveloped capacity for hydropower generation, an attractive proposition in a country where energy demands for manufacturing and household consumption are escalating rapidly and where three-quarters of current electricity supply is met by coal-fired power plants. The southwest region, with its rugged topography and high-volume, glacier-fed rivers, is home to the major share of China's vast hydropower potential. Development plans involving central-government ministries, provincial-government authorities, and limited-liability hydropower-development corporations are moving forward rapidly; China's Twelfth Five-Year Plan for Economic Development, released in 2011, specifically recommends pushing forward with the development of dams in the region, which is home to three of the country's thirteen so-called hydropower bases (*shuidian jidi*), areas targeted for the construction of large, electricity-producing dams.

Yunnan Province contains the upper reaches of five major river systems—the Pearl, the Jinsha, the Lancang, the Nu, and the Irrawaddy—which collectively have more than 600 tributaries and contain 221 billion cubic meters of water (Ma Jun 2004:178). This book focuses on two watersheds in Yunnan Province: the Lancang River and the Nu River. On the Lancang, where more than a dozen dams are planned, four are completely operational, and several are under construction. On the Nu, a thirteen-dam hydropower-development plan is under way, with a total hydropower potential of 21,000 megawatts, which is slightly more than the mammoth Three Gorges Dam. Should all thirteen dams in the cascade be built, the best estimates suggest that more than 50,000 people will be displaced. The effects of these dams on the environment and on the people who live in the region are immense but as yet poorly understood by scientists, policy makers, and the general public.

As a tool for economic development, dam construction is certainly not new on the scene. In fact, the world is home to more than 50,000 large dams, which the International Commission on Large Dams (ICOLD) defines as those greater than 15 meters in height or having a storage capacity greater than 3 million cubic meters (Scudder 2005:2–3). But China's role in this trend is startling: home to half of the world's large dams, it has far outpaced all other countries in the construction of dams in the past several decades and adds dozens of dams to its portfo-lio each year. The benefits provided by such projects are considerable: dams deliver hydropower, provide reliable irrigation water, enhance the navigability of waterways, and protect people and farmland against flood-ing. As hydropower meets a larger share of energy demand, it may also help to reduce the consumption of fossil fuels; government agencies and private entities alike are pursuing alternative-energy-development plans involving not just hydropower but also wind, solar, wave, and biogas. The development of a so-called low-carbon economy (*ditan jingji*) is wel-come news in a country where hundreds of thousands of people die each year from ailments linked to air pollution from fossil-fuel combustion (Economy 2004) and where pollution-related economic losses cut into the nation's gross domestic product (GDP). Given that China surpassed the United States in 2007 to become the world's largest emitter of green-house gases, such policy decisions also have global implications in the push to mobilize technology and political will to address climate change.

But dams also have consequences for ecosystems that are likely irre-versible, and many such consequences have long gone unaccounted for. The World Commission on Dams (WCD), an organization under the guidance of the World Bank and the World Conservation Union, pub-lished a landmark study in 2000 that concluded that although dams had contributed significantly to human development over the years, their deleterious effects on social and environmental systems had eluded meaningful scrutiny: "Dams have made an important and significant contribution to human development, and benefits derived from them have been considerable. . . . In too many cases an unacceptable and often unnecessary price has been paid to secure those benefits, espe-cially in social and environmental terms, by people displaced, by com-munities downstream, by taxpayers, and by the natural environment" (WCD 2000:6).

When a new hydropower dam is installed on the main stem of a major river, it fragments the riparian ecosystem, changing a free-flowing river segment into an expanse of still water. In the process, it disrupts fish pas-sage; alters the water's temperature, chemistry, and sediment load; and changes the geomorphology of the river itself, often in ways that are diffi-cult to predict. There is mounting evidence that the enormous weight of large reservoirs may even disrupt tectonic plates, a phenomenon called "reservoir-induced seismicity"; some experts speculate that Sichuan's catastrophic 8.0-magnitude Wenchuan earthquake, which killed an estimated 80,000 people in 2008, was caused by pressure on the earth's crust from the recently completed Zipingpu Dam, located within a kilometer of the quake's epicenter (Klose 2012). Moreover, critics of the dam industry point out that the heavy sediment load of southwest China's rivers, trapped in the still water of a reservoir, will build up so rapidly that the viable lifespan of any dam is not likely to exceed fifty years.

As an anthropologist, I have focused over the past few years mainly on understanding the human consequences of hydropower development, which are equally dire. When dams are built and reservoirs fill behind them, they displace the human beings who live there, flooding farmland, inundating homes, and changing lives forever. The effects can last for generations as people cope with the consequences for their family's income, their way of life, and their sense of place and community.[1] China's growing hydropower sector thus represents one of the key arenas in which the competing rationalities of economic development, energy production, biological conservation, and social welfare collide. Dai Qing, the well-known journalist and environmental activist, made a pertinent comment about the mammoth Three Gorges Project that aptly describes environmentalists' views of hydropower more generally: "The government built a dam but destroyed a river" (quoted in Watts 2011).

My goal in this book is to use water-resource management and the current drive for hydropower development as points of entry into an examination of the difficult choices faced by Chinese leaders about how to meet the nation's escalating energy demands without exacerbating the country's social and environmental problems. The massive dam projects under way in Yunnan, along with scores of others on most of China's major river systems, highlight the fact that water is simultaneously a resource that is central to people's livelihoods, a kinetic force capable of producing renewable energy to sustain national development, and a medium through which social and political relations are negotiated.

This path of inquiry opens up many questions that have thus far gone unexamined. What are the values and goals of key constituent groups in water-resource management in China, including government agencies, hydropower corporations, conservation organizations, and local communities? What strategies do these groups use to participate in the decision-making process and steer it toward the outcomes they deem desirable?

How do communities uprooted by dam construction and resettlement cope with the dramatic social, cultural, and economic changes they face? How do those in positions of official power balance the social and ecological costs of hydropower development against other imperatives such as energy security and integrated river-basin management? I use two main analytical concepts—statemaking and the moral economy—to examine these interrelated questions. In what follows, I briefly introduce these concepts and discuss how they help elucidate the current controversy surrounding hydropower development.

STATEMAKING AND DEVELOPMENT IN SOUTHWEST CHINA

Despite its location in the remote southwest, Yunnan has been influenced by Han Chinese dynastic expansion for centuries if not millennia. Indeed, peripheral places such as these, endowed with the natural resources required by a growing empire, have played key roles in the construction and maintenance of the Chinese state. This process of contact between core areas and peripheral areas—sometimes peaceful, sometimes conflictive—can be seen as part of the process of statemaking (Sivaramakrishnan 1999; Scott 1998). As James Scott observes in his seminal book *Seeing Like a State*, "Contemporary development schemes . . . require the creation of state spaces where the government can reconfigure the society and economy of those who are to be 'developed'" (1998:185). A growing body of work on the policies and politics of statemaking encourages us to examine critically how states are built and constantly shaped through geopolitical, cross-cultural, and sometimes discursive negotiation (Michaud 2010; Sivaramakrishnan 1999).

In this sense, statemaking is a development path characterized by modernist ambitions; it entails an essential faith in the power of human knowledge systems—including science, technology, and policy—to beget better and better futures. In the West, modernism's high-water mark was perhaps the middle of the twentieth century, when many of the benefits of scientific progress were realized but before the critical, reflective trend of postmodernism took hold. By contrast, Chinese states have been modernist in their tendencies since at least the Han Dynasty (206 B.C.E.–220 C.E.), with its trade routes into Central Asia, its uniform system of weights and measurements, its finely tuned legal institutions, and its civil service examination system that selected political leaders on the basis of academic mastery.

The role of Yunnan in the Chinese statemaking project is somewhat paradoxical. One need only glance at a historical chart of the dynasties, stretching back 3,000 years, or look at the way that the geographical extent of the dynasties waxed and waned throughout that long expanse of history to see that in the context of the modern Chinese state Yunnan can reasonably be considered part of the southwestern periphery. Yunnan became a province officially only in the late thirteenth century under the Yuan Dynasty of the Mongols. In fact, if we shift our gaze a bit, we can also see Yunnan as the northern end of the Southeast Asian highlands, a region called "Zomia," a name that scholars use to describe the vast upland zone that includes parts of southwest China, Myanmar, Thailand, Laos, Cambodia, and Vietnam (Scott 2009; van Schendel 2002).[2] These regions, which comprise some of the most mountainous terrain on earth and support a population currently numbering nearly 200 million, have long posed a challenge to the various governments that have attempted to subjugate them. Subsistence strategies, political affiliations, and even ethnic identities have long remained fluid here; many of the region's inhabitants have pursued swidden agricultural cultivation and a "deliberate and reactive statelessness" (Scott 2009:x). Although it would be a mistake to see the people of Zomia as entirely lawless or stateless, their current experience is one of far-reaching social, cultural, and environmental change (Michaud and Forsyth 2011). In Yunnan, consolidated Chinese control has been continuously threatened and often explicitly undercut by regional fiefdoms, interethnic conflict, and even, in the mid–nineteenth century, a Hui Muslim insurrection marked by widespread bloodshed (Atwill 2005). The Chinese fought the Vietnamese in a border war over Yunnan in 1979, a simmering cartographical and geopolitical conflict that wasn't resolved until the early years of the twenty-first century. It is here, amid the competing processes of assimilation and autonomy, integration and resistance, where some of China's greatest contemporary social and environmental challenges can be most clearly seen.

How does statemaking take place?[3] In this book, I consider statemaking from both a material perspective and a normative perspective. In a material sense, the central question is how powerful actors—including

government agencies but also, in the case of hydropower development, corporations with licenses to build dams on and distribute electrical power from China's southwestern rivers—accomplish the goal
of national development by drawing upon the resources of peripheral
regions such as Yunnan for exploitation. The geographer Darrin Magee
(2006) has developed a useful analytical concept called a "powershed"
that helps us understand how the highly developed regions of eastern
China in general and the Pearl River delta in particular have grown affluent in part because of a series of policy decisions that extract electrical
power from southwest China's rivers and allocate it in the east where
commercial and manufacturing demand is highest. This reallocation
allows distant resources to be used in strategic ways that maximize their
benefit to national development. Such policies represent a continua-
tion of uneven development stretching back decades. Deng Xiaoping's
Reform and Opening (Gaige Kaifang) policies during the 1980s empha-
sized a coastal-development strategy to attract foreign trade and investment, while China's western regions, with limited access to markets and
poorly developed transportation infrastructure, fell farther behind the
fast-paced development of the east (Lai 2002).

Although China is currently the world's second-largest economy and
will likely overtake the current leader, the United States, within a decade
or two, its distribution of wealth is among the most inequitable in the
world. Standards of living in Beijing, Shanghai, and other cosmopolitan
cities are similar to those in the West, but in many parts of the rural
interior they are more on a par with sub-Saharan Africa: hundreds of
millions of people struggle to find gainful employment, obtain clean
drinking water, and gain access to basic sanitation services. This is a
common trend in reform-era China; measured a variety of ways, income
and wealth inequality have continued to grow throughout the Reform
and Opening period, widening gaps between rural and urban communities as well as between individuals within communities (Naughton
2006). This pattern of uneven development has been aptly characterized as "one country with four worlds": the high-income areas of the
eastern coastal region, including Beijing and Shanghai; the middle-
income areas such as Tianjin, Guangdong, Zhejiang, and Jiangsu, also
on the east coast; the low-income, primarily agricultural areas of China's
interior; and the remote western areas with substandard living conditions (Hu 2003).

Yunnan's position in this economic geography is not an enviable one. The United Nations Development Program (UNDP) (2008) uses the Human Development Index (HDI)—which includes a measure of economic productivity, life expectancy, and education—to identify development needs and gauge the effectiveness of development interventions that are designed to increase quality of life. In a recent calculation of the HDI for China, Yunnan ranks thirtieth out of thirty-one provincial-level units; only Guizhou, its provincial neighbor to the east, scores more poorly.

Although statemaking can be seen as a material project, its normative dimensions—those driven by human values and perceptions—are equally important. How do the center and the periphery engage in a dialogue, each emerging with characteristics shaped by its interactions with the other? Ethnic identity, constructed over centuries of center–periphery interactions, is a crucial part of this story. The preamble to the Constitution of the People's Republic of China (PRC) explicitly declares the country to be a "unified, multiethnic state [*tongyi duo minzu guojia*]." In addition to the dominant Han majority, which constitutes about 93 percent of the nation's population, there are fifty-five "minority nationalities" (*shaoshu minzu*) that received formal recognition by the central government following an ethnic-identification project (*minzu shibie*) conducted between 1950 and 1956. The minority-nationality population, which numbered 114 million as of the 2010 census, represents a special development problem for the government bureaucracy. On the one hand, their perceived cultural and economic "backwardness" provides normative justification for targeted development, economic assistance, educational subsidies, and national welfare policies. On the other hand, many perceive the high concentrations of minority nationalities as a barrier to actually achieving development at a level consistent with the national average.

The anthropologist Fei Xiaotong, who studied in Britain under Branislaw Malinowski, one of the early pioneers of the ethnographic research method, played a key role in the ethnic-identification project of the 1950s. The project's general aim was to classify each minority group according to Marx's schema as primitive, slave, feudal, bourgeois-capitalist, socialist, or Communist (Harrell 1995). The government borrowed its classificatory criteria from the Soviet Union's system, developed by Stalin, which included a common territory, language, mode of subsistence, and, more obliquely, "psychological makeup." In practice, the process of affording recognized minority status to some groups while denying it to others was highly politically charged because it was tied to the establishment of ethnic regional autonomy for governance in minority areas (Wang and Young 2006).

Various places with high concentrations of ethnic minorities have received the designation of "autonomous region" (*zizhi qu*), "autonomous prefecture" (*zizhi zhou*), or "autonomous county" (*zizhi xian*), but the ability of officials within these entities to practice self-governance or to influence central policy remains extremely limited (see Rossabi 2004). Tibet, for example, is recognized as a provincial-level autonomous region (*zizhi qu*), but the potential for a secessionist movement causes the central government to keep a tight rein on the region. Moreover, China has been reluctant to accord its minority nationalities "indigenous" status under such frameworks as the UN Convention on Biodiversity or the UN Declaration on the Rights of Indigenous Peoples—both of which promote the rights of indigenous people for autonomy and self-determination.[4]

Fei Xiaotong conducted much of his pathbreaking research on rural Chinese society in Yunnan. In a series of case studies published with his student Zhang Zhiyi, he commented that Yunnan "is not very accessible from the central provinces; and, since distance breeds suspicion, only yesterday the age-old belief was still current that Yunnan was a wild region overrun with beastlike aborigines" (Fei and Zhang 1945:7). Indeed, contemporary public discourse about minority people in China is still marked by a mixture of pride, condescension, and humor. Most regions with large minority populations today feature minority theme parks, usually called "ethnic villages" (*minzu cun*), where Han Chinese visitors can stroll back in time to observe actors wearing minority costumes and posing in traditional houses or making handicrafts or singing and dancing to ethnic tunes.

In Dali, the capital of the ancient kingdom of Nanzhao, which clashed with both Tibetan forces and Tang Dynasty troops throughout the eighth and ninth centuries and which is now a prominent tourist destination, I recently saw a rhyming poem inscribed on the wall of a restaurant just outside the old town. The poem, entitled "Eighteen Oddities of Yunnan" (Yunnan shiba guai), resonates with many of the Han Chinese tourists who now flock to this region in search of ecological and cultural treasures.

It is not attributed to any particular author, and several slightly different versions now circulate on T-shirts and the Internet. It reads:

YUNNAN SHIBA GUAI
―――――――――――

Mazha dangzuo xia jiu cai
Caomao dang guo gai
Xiezi houmian duo yi kuai
Jidan shuanzhe mai
Fangzi kong zhong gai
Niunai zuo cheng pian pian mai
Si ji yifu tong chuan dai
Huoche meiyou qiche kuai
Baba jiao er kuai
Doufu shaozhe mai
Zhutong dang yandai
Guniang beizhe wawa tan lian'ai
Xian hua dang shucai
Shan ding tou zai yun tian wai
Yige liusuo guo jiang lai
Shui you qingqin huo you ai
Shantou dingzhe yi ge hai
Lao tai pa shan bi hou kuai

EIGHTEEN ODDITIES OF YUNNAN
―――――――――――――――

Grasshoppers are served as a snack while drinking
A straw hat can be a pot cover for cooking
Shoes have a flap on the back for hanging
Eggs are sold in rustic straw wrapping
A house is no more than a tarp for roofing
Milk is sold by the sheet after curdling
All four seasons require the same clothing
The train is slower than a car for traveling
Flat rice noodles are great for snacking
Tofu is sold after roasting
A length of bamboo makes a good pipe for smoking

Young women get married while their first born lugging
Fresh flowers count as a vegetable serving
Mountaintops appear above the clouds peeking
A zipline is the only means for river crossing
Water and fire are imbued with feelings
The mountaintop wears an ocean's draping
Old women are quicker than monkeys at climbing

Although my translation undoubtedly lacks much of the cleverness of the original, which performs linguistic somersaults in order to end each line with the vowel sound *ai* in the Chinese word for "odd," *guai*, the gist of the poem is clear: Yunnan is a land of mountain peaks shrouded in clouds, whose human inhabitants, for better or worse, have yet to be tainted by modernity. Their foodways are strange. Their sexual mores are exotic. And their animistic spiritual beliefs are as untamed as the landscape that surrounds them. Ideas about minorities in China operate almost at the level of *doxa*: they are taken for granted, unquestioned concepts about "the natural order of things" (Bourdieu 2004, 1990). In the Chinese rendering of the word *culture, wenhua*, the second character, *hua*, is transitive: it means "to become," "to transform from one state to another," the implication being that a person with culture is one who adopts Han customs and becomes literate in Chinese.[5]

The so-called nationalities question (*minzu wenti*)—how to integrate minorities into the national culture and economy in order to avoid social discord—has been a subject of political concern for centuries. Shortly after the defeat of the Nationalist Army and the establishment of the PRC, Deng Xiaoping, who three decades later would lead the country through the liberal economic transformation known as Reform and Opening, addressed a group of committed Chinese Communist Party (CCP) officials. The nationalities question figured prominently in his speech: "If we cannot resolve the nationalities problems, we cannot resolve the national defense problem. . . . Particularly in regards to the southwest region, we ought to make the nationalities a high priority" (Deng [1950] 2006:194). Ever the pragmatist, Deng saw the best solution as one that maintained a tradition of political cadres drawn from the ranks of local minority people so long as these local leaders remained loyal and subordinate to the party.

The nationalities question continues to ignite passionate debates in scholarly and policy circles today. One recent example is an article by the Beijing-based social scientists Hu Angang and Hu Lianhe (whom scholars playfully refer to as "Er Hu" after the two-stringed instrument in classical Chinese music) published in the journal *Sociology of Ethnicity* and titled "Second-Generation Nationalities Policy: Promoting Ethnic Integration and Prosperity" (2011). They argue for an assimilationist approach that would entail dramatically scaling back social programs for minorities, including preferential admission to educational institutions, and even removing altogether the category of *minzu* from the national identification card. Citing the social, economic, and security benefits that stem from being a part of the Chinese nation-state, they argue for national identity to take its transcendent place above more narrowly defined regional or ethnic affiliations. Their views have raised howls of protest from various scholars, who see this argument as wildly optimistic, if not downright naive.[6]

The normative aspects of statemaking have important implications for the management of natural resources, including water. Conservation-oriented nongovernmental organizations (NGOs), which fight an uphill battle in China, often resort to strategic essentialism to promote the idea that the *minzu* who reside in a particular area are "ecologically noble savages" with a natural inclination toward responsible resource management if not explicit conservation. In the Tibetan areas of northwest Yunnan, for example, conservation organizations—both international, such as Conservation International, and domestic, such as the Kawakharbo Society—promote this notion in order to limit hunting, logging, or the gathering of nontimber forest products. These organizations cite the core principles of Buddhism, which is deeply ingrained in Tibetan culture, invoking doctrinal themes such as karma and the ideal of not harming sentient beings to suggest that the region's minority people are natural conservationists (Yeh 2014).

The central government and other proponents of dam construction are no less adept at using strategic essentialism, but their position, articulated in policy documents and government-sponsored academic publications, is precisely the opposite. Government agencies regularly depict the practices of highland *minzu*, including swidden agriculture and the unregulated harvest of nontimber forest products such as mushrooms and herbs, as harmful and exploitative of the environment. This depiction buttresses their overall argument, which is that the displacement and resettlement of communities for dam construction, however challenging and riven with conflict, will ultimately benefit the environment by protecting it from exploitation at the hands of irrational users. Moreover, dam proponents explicitly argue that hydropower development constitutes a much-needed poverty-alleviation strategy in an economically backward region that lacks jobs and other forms of investment (He 2009).

THE MORAL ECONOMY

A second analytical concept that shapes the narrative of this book is the idea of the moral economy, which can be understood as "popular consensus . . . grounded upon a consistent traditional view of social norms and obligations, of the proper economic functions of several parties within the community" (Thompson 1971:79). The notion goes back to a seminal work by historian E. P. Thompson entitled "The Moral Economy of the English Crowd in the 18th Century" (1971). Its central premise is that contemporary economic and political interactions are shaped by norms with much deeper cultural and historical roots. Anthropologists and their colleagues in related social science disciplines have long been interested in the underlying concepts and motivations, both explicit and tacit, that drive people's behavior and interactions with others. In the most fundamental sense, morality—from the Latin *moralitas*, meaning "proper behavior"—is the socially negotiated division of actions into categories such as good and bad, proper and improper.[7]

In a retrospective essay twenty years after the publication of "The Moral Economy of the English Crowd," Thompson clarified that his version of the moral economy was "confined to confrontations in the market-place over access (or entitlement) to 'necessities'" such as food, whose distribution in the political economy of the marketplace could be uneven and unjust (Thompson 1991, quoted in Edelman 2005:331). However, since its formulation, the moral economy concept has been expanded, stretched, and scaled up to describe wider and broader social phenomena. Scholars have used it to examine people's ideas about rights to subsistence and agricultural production (e.g., Scott 1976); expectations of the state in the provision of social services (e.g., Sivaramakrishnan 2005); the role of market economies in producing and distributing commodities (e.g., Griffith 2009); and competing goals and priorities among water-management agencies, communities, and user groups (e.g., Wutich 2011; Mosse 2003).

Because of its central role in sustaining all other economic and cultural activities, water is a commodity that lends itself well to this kind of analysis. A society's political and economic choices about water use and allocation reveal a great deal about its core values and about how those values are constructed and maintained in everyday practice. Historical analysis suggests that when new economic, political, or environmental events violate people's sense of what is just or proper, civil unrest or mass social movements may result (Edelman 2005), which underscores the fact that a shared sense of propriety is at the core of human social relations. In reform-era China, the moral economy idea is a particularly useful concept because the country is currently undergoing far-reaching social and economic transformations that are pushing the boundaries of traditional social norms. As the CCP has liberalized the nation's economy over the past three decades, it has also gradually reduced the scope of its administrative power and thus increased the space within which civil society organizations may operate (Tilt 2010; O'Brien and Li 2007; Weller 2005). This is especially important in the environmental arena, where the new Environmental Impact Assessment Law (Chinese National People's Congress 2002a), promulgated in 2003, mandates public hearings for major development projects. Scholars point out that environmental lawsuits are increasingly common in China, as are environmental NGOs, which number in the hundreds (G. Yang 2005).

In short, different constituent groups argue for water-resource-management objectives that stem from different moral visions about how rivers should be managed. In using the moral economy as an analytical framework, I intend to examine the values, goals, and objectives of key constituent groups in water management, including government agencies, hydropower corporations, NGOs, and local communities. I explore how these groups' values are grounded in cultural norms about what is right or just or proper and for whom. I also examine the strategies that such groups use to influence water-resource decision making—including policy advocacy, law, public protest, and other means—and explore the outcomes associated with such strategies.

In China's water-management sector, different constituent groups use different moral arguments to support their positions about hydropower development. For the central government and for the large hydropower corporations that build and operate the dams, providing reliable electrical power to fuel China's economic growth is the primary concern. For local communities, the paramount moral issues center on people's ability to participate in the decision-making processes that have the potential to change their lives dramatically, to retain access to farmland, and to receive adequate compensation for their losses. For international conservation organizations, the moral imperative is to preserve a region of immense biodiversity; these groups effectively stake a claim on local resources as a form of global biological heritage.

Such moral concerns are commonly confronted in the arena of international development, but they take on greater complexity and nuance in the context of contemporary China, which for the past several decades has undergone the most far-reaching economic transformation in history. Development remains a powerful concept in the modern world, one that mobilizes a great deal of capital and expertise. In the undergraduate course I teach on international development, I often begin the first class lecture by telling students that we will examine the concept of development through three interrelated lenses: its objectives, its instruments, and its outcomes. The objectives of development—infrastructural improvement, poverty alleviation, the expansion of educational opportunities, the reduction of mortality and morbidity, among others—have been around for a long time, even as they have undergone significant evolution. The international financial institutions created by the Bretton Woods Accords in 1944, including the World Bank and the International Monetary Fund, were designed to rebuild Europe after World War II; after accomplishing this goal, of course, they did not fade into obsolescence but rather refocused their energies and retooled their agendas to address poverty alleviation in the developing world.

However, the instruments of international development vary considerably across regions and across time periods. Should national governments take the lead, or should multilateral agencies? What role should NGOs play? Many of the textbooks commonly used in international development courses fail to include much detail on China, probably because the authors have a difficult time placing the country into one of the categorical paradigms of development theory. China doesn't fit "modernization theory" as well as Japan does. It doesn't exemplify "dependency theory" like Chile or Argentina. Given the market-oriented reforms of the past several decades, it doesn't fit "state-led development" quite as well as it once did. Nor does it illustrate "neoclassical development" quite as effectively as Mexico. China's recent development path is a hybrid model, combining market logic, a continued legacy of state planning in key sectors, and close ties between political and economic elites. It is essentially state-managed capitalism, although leaders much prefer the euphemism "socialism with Chinese characteristics" (*Zhongguo tese de shehui zhuyi*).

The outcomes associated with China's development path over the past few decades are nothing short of remarkable: its coastal cities have grown into global manufacturing centers; its countryside has seen the dismantling of collective agriculture in favor of small-scale, lease-based, private farming; and hundreds of millions of people have joined the ranks of the middle class. In this regard, no matter how we choose to classify China's development path, it shares one crucial characteristic with many Western models: the idea that development is an eminently desirable pursuit—one that is politically neutral, driven forward by faith in scientific, technological, and economic progress (Ferguson 1994). As demonstrated by many of China's largest engineering projects, including the dams that constitute the main focus of this book, the CCP's overriding philosophy is one of pragmatism. It seems there is no problem for which a technocratic solution cannot be found.

Despite the fact that China and the West share crucial common ground on the issue of modernism, understanding China's recent development path requires a slightly different way of thinking, not because the market fails to operate here, but because in addition to that "invisible hand" there is also the constant, guiding hand of the party–state. Anna Tsing has referred to this melding of market and authoritarian state as the "frontiers of capitalism" (2005:27). The melding process is not always easy and entails the same philosophical battles that have played out for centuries between liberal economic theorists, who wish to see as little regulation as possible on rational economic actors in the marketplace, and Keynesians, who see a key role for the state in coordination, planning, and the provision of basic social services. At the moment, China seems to be having its cake and eating it, too. As I argue in this book, the convergence of capitalist development with the party–state's stronghold on political power is an incredibly effective way to fast-track large hydropower projects. It can also spell great hardship for individuals and communities who stand in the path of development but lack a seat at the decision-making table.[8]

Within the moral economy of water resources, questions of expertise—of the different ways of constructing and valuing knowledge—are central to this discussion. How do the main constituent groups in water-resource management and hydropower development—government agencies, hydropower corporations, NGOs, and local communities—conceive of the dam-development issue, produce knowledge about it through the use of science and other means, and seek to influence the decision-making processes surrounding it? These groups have quite divergent perspectives on water, with some viewing it as a global resource to be conserved, some as a resource to be developed for the good of the nation, and some as a critical component of a landscape in which they live. The dams that constitute the major focus of this book have been built—or are currently being built—in the name of national economic development. But the normative questions that surround them—about transparency and public participation, about the balance between conservation and development, and about the obligations of government agencies to protect vulnerable individuals and communities—remain unresolved.

A TALE OF TWO WATERSHEDS

Subsequent chapters of this book provide detailed case studies of two watersheds in Yunnan Province where intensive hydropower development is under way: the Lancang River and the Nu River. These case studies provide a ground-level examination of the cultural, ecological, and economic impacts of hydropower development for local communities. I wish to provide just enough information here so that readers will have some context for understanding these two watersheds and their role in the narrative of this book.

In one sense, the choice of these two river basins for detailed analysis is somewhat arbitrary; I could just have easily have chosen to concentrate on development schemes on the next major watershed to the east, the Jinsha, or on the Dadu or the Yalong or on a host of other rivers where dams can be seen rising, a ton of concrete at a time, over canyon walls.[9] In another sense, however, the Lancang and the Nu are natural choices for a study of this sort, for several reasons. First, key government policies, including the Twelfth Five-Year Plan for Economic Development (2011–2015), single out the southwest region as a major base (*jidi*) of hydropower development, and these two rivers offer a glimpse into two very different stages of development. The Lancang has seen fairly rapid hydropower development, whereas the dam projects on the Nu are proceeding more slowly and with significant international and domestic contestation. Second, as I have already suggested, these two cases involve particularly high stakes for biological diversity, cultural heritage, and even international relations with downstream countries. Finally, and perhaps most significantly, my involvement since 2006 with a large, interdisciplinary project designed to understand and model the potential impacts of dams has inspired much of my interest in this topic, and these two watersheds have been the focus of that effort from the beginning.

These two rivers traverse an area of truly remarkable biological and cultural diversity, from the glacier-covered peaks of Khawa Karpo, which towers 6,740 meters above the provincial boundary between the Tibet Autonomous Region and Yunnan, to the subtropical rice terraces and rubber plantations of Xishuangbanna, which abuts Laos and Myanmar. The cultural practices and livelihood strategies of the people who live here and who identify with more than a dozen *minzu* are also tremendously diverse. The landscape itself is freighted with epistemological significance: to the people who live here it is a working landscape with cultural and even spiritual significance; to Han Chinese and foreign visitors it is the mythical and exotic Shangri-La; and to Western scientists and conservation groups it is a biodiversity hot spot renowned for its flora and fauna and in need of protection. Since the late nineteenth century, Western naturalists, missionaries, and explorers—sometimes collecting plant specimens for scientific analysis or conservation, sometimes spreading religion, always with a great talent for "captivating writing and self-promotion" (Harrell 2011:3)—sent back word of this wonderland to eager readers in Europe and the United States.[10] Their writings in both scholarly journals and popular texts have played an important role in shaping the public imagination of Yunnan.

Fei Xiaotong and Zhang Zhiyi's *Earthbound China* (1945) is a classic study of the Chinese rural economy, based on fieldwork in villages scattered to the west and south of Kunming, the provincial capital of Yunnan.

Undertaken when the authors were at National Yunnan University in the years leading up to World War II, it argues that Yunnan was illustrative of what much of rural Chinese society was like for centuries: "Since in interior China modern industrial and commercial influence is just beginning to be felt, village folk are still farming with the old techniques, economically more or less self-sufficient, and are imbued with the traditional virtue of contentment. The population is dense, and resources are limited. It is old China in miniature" (vii).

Figure 1.1 shows a map of both river basins. The Lancang River originates at 5,200 meters high on the Qinghai–Tibet Plateau, then winds its way for 2,200 kilometers through Yunnan Province before flowing, as the Mekong, through five downstream riparian nations in Southeast Asia, where it helps to support the livelihoods of nearly 60 million people. The Lancang basin is undergoing large-scale hydropower development in two phases, referred to as the Lower Cascade and the Upper Cascade. The development plan for the Lower Cascade, which is already quite far along, calls for a total of seven dams, four of which—the Manwan, Dachaoshan, Xiaowan, and Jinghong—are now operational (Magee 2011; Dore, Yu, and Li 2007). Their designs and operation plans vary considerably: whereas Manwan and Jinghong are comparatively small with limited reservoirs, Xiaowan and Nuozhadu are taller than the Three Gorges Dam and have already displaced tens of thousands of villagers. The Upper Cascade development plan, which is still under revision, calls for anywhere between five and twelve dams in the primarily Tibetan areas of northwest Yunnan (Magee 2011); it remains speculative in part because of the excessive cost of building roads and other associated infrastructure in an extremely underdeveloped region, although recent media reports confirm that preparatory work is under way in at least two locations, Lidi and Wunonglong.

The Nu River originates at more than 5,500 meters high on the Qinghai–Tibet Plateau, before flowing southward through Yunnan and continuing through Myanmar, where it forms part of the Thailand–Myanmar border and is known as the Salween. With a total length of 2,018 kilometers, the Nu is often referred to as "Asia's longest free-flowing river," although several dams are under construction downstream in Myanmar, with financial backing from Thai and Chinese government agencies and hydropower corporations. The name "Nu," which translates literally as "Angry," is actually a phonetic rendering of "Nong," the name given to the river by the local Lisu ethnic group (Mertha 2008).

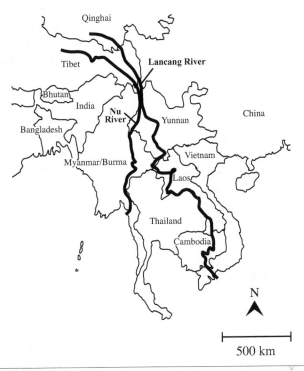

FIGURE 1.1 Map of the Lancang and Nu River courses in China and mainland Southeast Asia.

But it is also meant to convey a sense of the turbulent nature of the river, an apt metaphor for the controversy surrounding the proposed dam-development project. A thirteen-dam hydropower-development plan is currently under way, with a total hydropower potential of 21,000 megawatts, which is slightly more than the mammoth Three Gorges Dam. Should all thirteen dams in the cascade be built, the best estimates suggest that more than 50,000 people will be displaced. As subsequent chapters demonstrate, the Nu River dam projects have spurred domestic activism as well as highly public international criticism from conservation organizations. Development plans have been started, stalled, and restarted nearly half a dozen times over the past decade.

RESEARCH METHODS AND "THE FIELD"

For a researcher like me, an anthropologist with a track record of working on rural development issues, "the field" is usually conceptualized as someplace far-flung or exotic. This is the common supposition about what it means to conduct anthropological fieldwork, and it has shaped my discipline in very powerful ways. Fieldwork in Yunnan over the years—particularly in places involving tedious and uncomfortable travel—has certainly fit that bill. Material for this book was collected during field visits in 2006, 2008, 2009, 2010, 2012, and 2013 involving both personal and collaborative research through interviews and household surveys with villagers whose lives are being upended by the dam projects on the Lancang and Nu Rivers.

But this particular research project is substantially different from what I have become accustomed to over the years. Rather than spend a long period of residence in a single community—a method of inquiry that anthropologists call "participant observation"—I collaborated with other scientists to collect survey data from more than 1,200 households in dozens of villages across these two watersheds. This approach is indicative of a broader trend in anthropology and related disciplines toward holism and cross-disciplinary collaboration that effectively reenvisions what it means to go into "the field" (Gupta and Ferguson 1997).

Furthermore, my research in rural villages constitutes only part of the story of this book. I also spent six months in Beijing as a Fulbright scholar in 2012, an opportunity that opened many doors to better understand how management decisions about water resources are made at high institutional levels. My scholarly insights in this area have as much to do with serendipity as with careful planning. While I was preparing for research in 2012, forces beyond my control caused one delay after another in my affiliating with an institution and securing the right type of travel visa; these delays ultimately convinced me to affiliate with colleagues in Beijing and to alter my project accordingly.

As a result, in additional to rural villagers, my research "subjects" came to include engineers, scientists, policy makers, NGO representatives, and other experts who play a role in water-resource management and hydropower development. I conducted dozens of interviews with experts involved in research and regulatory oversight on the environmental and social dimensions of hydropower. In the process, I came to reenvision the field in a more inclusive way that in addition to rural villages encompassed academic institutions, government offices, and consulting firms located in the glass-and-steel high-rises of modern Beijing.

I discovered along the way that although water may not flow from Beijing—the North China Plain aquifer is, after all, perilously depleted—power most certainly does. Despite recent trends toward the decentralization of water-resource management in China, government agencies and hydropower corporations with close governmental ties play a prominent role in shaping the course of policy. Very little is known about the machinations of scientific expertise, knowledge, and influence within these organizations, and one contribution of this book therefore is to delve into such institutional processes. Of course, experts are people, too—equipped with histories, cultural biases, and strategic positions shaped by the political economy of the institutional environments in which they work—which makes them interesting subjects of analysis in their own right. That is not to say that I was given unrestricted access to the ins and outs of water-resource policy, which is still something of a sensitive topic in China, particularly on the transboundary rivers of the southwest, where even basic information about flow rates can be considered *neibu*, "for internal use only." Indeed, although I found most scientists to be open and enthusiastic participants in the study, I often struggled to gain inroads with policy makers and others with formal institutional power.

I have also conducted, with the help of students and colleagues, a relatively thorough review of documents such as comprehensive river-basin-management plans; environmental impact assessments (EIAs); corporate publications from China's "Five Energy Giants" (Wu da Fadian Jutou), the key companies involved in hydropower development; and law and policy documents from agencies such as the State Council, the Ministry of Water Resources (MWR), and the National Development and Reform Commission (NDRC). These documents supplement the data from surveys, interviews, and observations and help to provide a more complete picture of the complex array of actors involved in water-resource management, conservation, and hydropower development in Yunnan.

CHAPTER OVERVIEW

In chapter 2, I trace China's recent economic rise, its escalating energy demands, and the various forms of renewable energy currently being pursued by policy makers. Hydropower plays an increasingly important and conspicuous role in the national energy portfolio as China attempts to wean itself from fossil fuels. Although China shares this pursuit in common with many other countries, including the United States, the scale and pace of its hydropower-development scheme is truly unprecedented. The chapter concludes with a discussion of recent policy changes that have transferred the construction and operation of dams and the distribution of electrical power from government agencies to private investors or shareholder corporations.

Chapter 3 provides a detailed case study on how dams are affecting the lives of people in the Lancang River basin, where four large dams have been completed and more are under way. I examine the effects of these projects on people's access to land, their agricultural practices, their household finances, their sense of place, and their social networks. An important part of this analysis is the complex topic of land rights in rural China, which have changed considerably in recent years. Some resettled villagers cope with displacement by farming new land plots and changing their cultivation and marketing strategies accordingly, whereas others have entered the wage-labor market or attempted to start their own businesses. Many villagers face a pattern of indebtedness that will pose a challenge for years to come.

In chapter 4, I turn to an analysis of the Nu River basin, exploring the current livelihood strategies of the people who live there, many of whom are among the poorest in China. I also examine the vulnerability of Nu River villagers to the proposed hydropower-development plan, addressing their precarious economic situation, their attempts to participate in the political decisions that affect their lives, and their struggle for cultural autonomy. Despite the fact that the Nu River dams have received extensive international media attention, many local villagers have little specific knowledge of the construction plans. Their views of hydropower development are characterized by great uncertainty, tempered with the optimistic expectation that government compensation programs will bring them some benefits and help them to avoid some of the worst social problems associated with displacement.

Chapter 5 focuses on recent collaborative efforts across scientific disciplines to understand and mitigate the complex problems associated with dams. How do scientists and policy makers use data to reach decisions about the management of water resources? I address this question in two ways: first, by drawing upon observations and interviews with scientists and policy makers within Chinese government agencies and NGOs; and, second, by reflecting on my own experience as a researcher on a large, interdisciplinary project designed to create a decision-support model that helps policy makers predict, visualize, and weigh the many costs and benefits of dams. A significant part of this story relates to how science operates as an epistemological domain, how various scientific disciplines construct knowledge in disparate ways, and how these various forms of knowledge come to influence the policy-making process.

In chapter 6, I address the most pressing social problem associated with dams: the uprooting of communities through forced displacement and resettlement. I discuss China's policy framework for resettlement compensation, which has improved considerably in recent years but remains problematic and unevenly implemented. I also discuss two primary tools used by social scientists for mitigating the effects of dams on communities: social impact assessment and participatory development. Both of these tools hinge upon a notion of individual rights, which calls for a consideration of the legal, political, and cultural barriers to a rights-based framework in China. I consider evidence from the Lancang dam projects that shows a trajectory of improvement in the implementation of resettlement and compensation policies over recent decades.

I widen the analytical lens of the book in chapter 7 to consider two global processes related to hydropower development: first, the increasingly important role of international NGOs operating in China, which advocate for environmental conservation and cultural preservation in Yunnan; and, second, the growing role of Chinese government agencies and corporations in providing financial and technical assistance to dam projects in the developing world, especially Africa and Southeast Asia. The linkages between China and other global players in dam development have dramatically changed the way that hydropower projects throughout the world are planned and implemented, from a model in which international financial institutions such as the World Bank took the lead to one in which the Chinese dam-building industry has risen to prominence. This trend has important implications for assessing the future of dams as a development mechanism around the globe.

In the final chapter, I revisit the moral economy concept to examine how hydropower development has brought to the forefront a set of normative questions about how to balance economic development, alternative-energy production, and ecological and cultural concerns. Different constituent groups tend to take extreme positions, with government agencies and hydropower corporations calling for more dams and conservation NGOs trying to stop or stall the projects. But the truly difficult task, the one that must be undertaken, is to envision outcomes that promote energy production while also taking full stock of social and ecological costs and devising accountability mechanisms that require relevant parties to pay them. Achieving both goals will ultimately require greater transparency, increased scientific collaboration, and broader public participation in decision making.

2

CRISIS AND OPPORTUNITY

Water Resources and Dams in Contemporary China

THE METEORIC rise of the Chinese economy and with it the emergence of a robust middle class with higher living standards and escalating consuming power constitute parallel threads in one of the great economic stories of modern times. It is a story of exponential growth that has lifted hundreds of millions of people from poverty and transformed an agrarian nation into a manufacturing powerhouse within the space of a few decades. But economic growth is always constrained by biophysical limits, and water—a resource that sustains all economic and cultural life—now represents one crucial point at which this economic miracle appears to be running up against such limits. Like many developing nations in which the demand for water has begun to outstrip supply, China faces a water crisis.

This crisis relates to water availability and water quality, to be sure, but it also increasingly involves large-scale hydropower-development projects that are transforming China's rivers. In this sense, water is not merely a precious resource in its own right but also a source of kinetic energy that can be converted to electrical power and transported to fuel economic development in commercial and manufacturing centers located thousands of kilometers away. The transport of hydropower from west to east traverses political, ecological, geographical, and cultural boundaries. Although there are many precedents in Chinese history for such feats of hydraulic engineering, today's ambitious hydropower dam projects are on a scale that has no equal in human history.

Balancing the economic benefits of hydropower with the ecological and social costs it causes is a technical problem, but it is also a normative one.

It is a question of the values and worldviews that shape people's relationships with water and how all of these things change over time. In the current era of market-oriented reform, China's water-management sector is becoming increasingly fragmented and even privatized, a trend that entails a reconfiguration of the relationship between state, society, and market.

AN APPETITE FOR ENERGY

China's recent development path, which is characterized by rapid economic growth and a reliance on manufacturing for the export market, has created an ever-expanding demand for electricity. In the early years of socialism after the establishment of the PRC in 1949, China followed the Soviet model of development, with industrial production in the hands of the central government and agricultural production controlled by a network of rural collectives. With the death of Mao Zedong and the ascendance of Deng Xiaoping in the late 1970s, Chinese leaders began a series of liberal reform policies known collectively as Reform and Opening, which ushered in sweeping social and economic change. These reforms have drastically altered the lives and livelihoods of one-fifth of humanity (more than 1.3 billion people), as the nation has seen a return of smallholder agriculture under the Household Responsibility System (Jiating Lianchan Chengbao Zeren Zhi), the privatization of industry, greater integration into the global economy, and the rise of an urban consumer class. China's GDP has grown nearly 10 percent annually over the past thirty years. Its economy is expected to be the largest in the world, surpassing that of the United States, within the next two decades (Tilt and Young 2007).[1]

Chinese leaders face critical challenges in the years ahead, including a global economic slowdown that has scaled back demand for manufactured goods and rising social unrest from the widening chasm between rich and poor. However, having emphatically arrived on the global economic stage, China appears poised to stay for the long term. Nothing tells the story of the nation's economic rise better than the upward trend of energy consumption that has accompanied it. On the eve of Reform and Opening in 1978, China was a poor, agrarian country. Between 1980 and 2010, the nation's total annual energy consumption increased by a factor of five, to reach the equivalent of 3.25 billion tons of coal being burned each year. Even conservative projections suggest that China's energy consumption will continue to rise in the coming years, if at a slightly slower rate, as its economy matures.

This is one part of the global geographical picture of energy consumption, which is changing dramatically as new and dynamic economies emerge. In 1971, the Asia Pacific region accounted for just 15.3 percent of global energy consumption; by 2010, it accounted for 38.1 percent, taking shares from Europe and North America (Liu 2012:5). China's rapid rate of annual growth, coupled with its huge population, means that China accounts for a lion's share of the growth in energy consumption in the Asia Pacific region. In particular, access to reliable and affordable electricity has lifted countless millions of people out of poverty by sustaining manufacturing jobs, improving the nation's transportation network, and supporting China's integration into the global economy.

Energy production and consumption, of course, pose serious environmental challenges. When it comes to the relationship between energy and the environment, the sources of electricity generation are of crucial importance. But it can be difficult to trace the environmental costs of electricity generation, in part because each source has a unique set of advantages and disadvantages related to its environmental impacts, operational costs, and long-term sustainability. For example, some electricity-generating sources, such as coal-fired power plants, emit a majority of their greenhouse gases during operation. Others, such as nuclear power plants, emit almost no greenhouse gases during the process of generating electricity but entail significant carbon emissions during the mining of raw materials, the construction of facilities, and the transportation and disposal of waste.

A recent assessment of greenhouse gas emissions from different modes of electricity generation around the globe used a life-cycle approach, which accounts for emissions from all phases of a given project and normalizes these emissions by calculating greenhouse gases per unit of electricity produced. Emissions can be expressed in a simple equation: CO_2e/GWh, or the weight of carbon dioxide equivalent per gigawatt hour of electricity. This gives us a measure of the intensity of greenhouse gas emissions, which is a key focus of the Intergovernmental Panel on Climate Change and nations around the world as global attention focuses on carbon emissions as a driver of climate change.

Based on this calculation method, it is clear that all electricity sources are not created equal. Coal-fired power plants produce, on average, 888 tons of CO_2e per GWh; natural gas produces about half that figure. Hydroelectric facilities produce an average of 26 tons of CO_2e per GWh, about one-thirtieth the level of coal-fired power plants, making hydropower comparable to renewable electricity sources such as solar and wind power (World Nuclear Association 2013).[2] This underscores the fact that Chinese leaders, like their counterparts in developed and less-developed countries alike, are often forced to choose between a set of bad options when it comes to electricity production.[3] Approximately two-thirds of China's current electricity demands are met by coal-fired power plants (see figure 2.1). China remains the world's leading consumer of coal, using approximately 1.5 billion metric tons per year for industrial production, energy generation, and, particularly in rural areas, household heating. The nation's coal stores are vast: at least 100 billion metric tons, enough to continue current rates of coal combustion throughout the twenty-first century (Smil 2004).

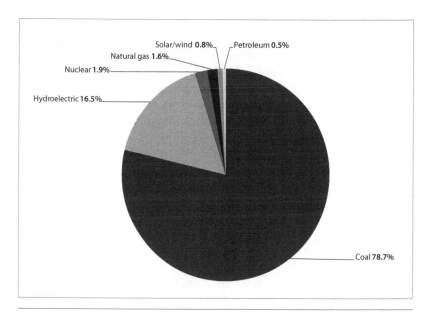

FIGURE 2.1 China's national electricity generation by source. *Source*: Liu 2012.

Although coal remains an attractive economic option in the near term, the environmental costs of coal and other fossil fuels have become more and more apparent. For the past several years, the U.S. embassy has been releasing its own data on Beijing's notoriously bad air pollution, which officials have collected from monitoring equipment within the embassy compound. When I resided in Beijing in 2012, air-quality ratings based on particulate matter typically hovered in a range between "unhealthy," "very unhealthy," and "hazardous." For several weeks during the winter of 2013, air-quality ratings exceeded the upper bounds of the scale by a factor of two. These figures have often been at odds with those released by China's Ministry of Environmental Protection (MEP), and the controversy has received considerable coverage in the press (Moore 2011), prompting a great deal of public discussion about air pollution as well as regular updates on levels of particulate matter in daily weather reports. Not that most Beijing residents need detailed data to tell them when the air quality is bad. When buildings just a few hundred meters away are obscured in haze; when any amount of physical exertion causes wheezing in the lungs; or when what goes into people's handkerchiefs comes out black, there is not much debate about how horrible the urban air quality is.[4]

China now has the dubious distinction of being home to sixteen of the world's twenty most-polluted cities, costing hundreds of thousands of lives and mounting economic losses from health problems, lost worker productivity, and infrastructural damage (Economy 2007, 2004). Beyond its borders, there are significant global implications to China's economic rise. In 2007, China surpassed the United States to officially become the world's largest emitter of carbon dioxide (Vidal and Adam 2007).

These aggregate figures, although dismal, mask several important facts about China's economic rise and its consequences for the environment. First, China's per capita rate of energy consumption is still below the global average (Liu 2012:19). With a population of 1.3 billion serving as a denominator in the equation, China's per capita emissions of pollutants and greenhouse gases are on a par with many lower-middle-income countries; the United States, meanwhile, remains in first place in terms of per capita emissions. Moreover, central policy decisions that support investment in cleaner production have improved energy efficiency for many of the nation's worst polluters in large-scale industries such as iron and steel production (Managi and Kaneko 2009). Non-point-source pollution from automobiles is becoming a greater concern, a trend that will likely continue as car ownership becomes a hallmark of membership in the middle class in China.

Second, China's manufacturing sector is export oriented, geared toward serving consumer demand in global markets, primarily in the West. That means that low labor costs and lax environmental enforcement, long China's twin comparative advantages in the global economy, make it "rational" for foreign corporations and even small companies to locate their dirtiest facilities in China. Although pollution may be a daily fact of life for most Chinese citizens, the economic impetus comes largely from consumer demand in countries with a much higher standard of living. It is a global problem driven by the externalization of environmental costs from Western countries to developing nations, including China.

Finally, the distribution of energy consumption, both in geographic terms and in socioeconomic terms, is highly uneven. These days the standard of living in cosmopolitan cities such as Beijing and Shanghai is not appreciably different from that in New York or London. I wrote early drafts of this book in a 110-square-meter apartment located in an upper-middle-class residential complex in northwest Beijing. Like my neighbors, I had reliable hot water, a refrigerator, a washing machine, a high-speed Internet connection, and a television set that was probably larger than I really needed. I can't help comparing this standard of living to that in the Chinese countryside, home to more than 800 million people, where I have conducted most of my fieldwork on rural development and where per capita energy consumption is only about one-third that of the nation's cities. In rural Yunnan, for example, many villagers lack access to indoor plumbing and have only one or two electric lightbulbs by which to see at night. Many areas have "leapfrogged" previous generations of communications technology; for example, in some of the poorest areas of Yunnan, it is rare for a household to have a landline telephone connection, but ownership of mobile phones is ubiquitous.

Although its dependence on coal and other fossil fuels will persist for many years, China is investing heavily in alternatives, which policy makers refer to as "clean energy" (*qingjie nengyuan*), "green energy" (*lüse nengyuan*), and even "low-carbon energy" (*ditan nengyuan*).[5] China has committed to reduce carbon emissions to 40–45 percent below 2005 levels by 2020—not just in raw terms, but also in carbon intensity, which measures the carbon emissions required to produce one unit of GDP, from a current level of 2.7 tons per $10,000 of GDP to a target of 1.2 tons.

The struggle to lessen China's dependence on carbon-based fuels is a difficult road, marked by fits and starts and a fair amount of contention. In the international arena, Chinese leaders are reluctant to acquiesce to policy agreements that compromise the nation's unprecedented rates of economic growth, which have been in or near double digits since the early 1980s. For example, Chinese representatives to the 2009 UN Climate Change Conference in Copenhagen, Denmark, pledged deep cuts in carbon emissions over the coming decades. However, like the leaders of other rapidly developing nations—including the other so-called BRIC countries (Brazil, Russia, and India)—Chinese leaders dragged their feet during the conference, which ultimately failed to reach any binding agreements between the world's largest economies.[6]

Amid these seemingly intractable problems, there are reasons for optimism. The MEP announced its intention to focus on co-control, reigning in both emissions of key pollutants linked to adverse health outcomes and greenhouse gases that contribute to global warming. China leads the world in key areas related to alternative-energy development, including the production of solar panels. Carbon dioxide emissions in the energy sector have leveled off, due in part to more stringent policy and in part to improvements in production output per unit of energy from the closure of many small, inefficient, coal-fired power plants (Managi and Kaneko 2009).

Water resources—and their capacity to be turned into hydroelectricity—are a central part of this story. Government documents report with a measure of pride that hydroelectricity output, which currently accounts for 16 percent of the nation's electricity portfolio, grew at an annual rate of 12.9 percent throughout the Eleventh Five-Year Plan period (2005–2010) and will continue on a similar pace for the foreseeable future (Chinese State Council 2013). Now more than ever before, top leaders view water not simply as a critical resource in its own right but as an important means for the clean production of renewable energy. All of these factors place greater strain on China's water resources, which are limited to begin with and which are not always distributed in ways that match demand.

WATER IN GLOBAL CONTEXT

It is difficult to imagine a more precious resource or one more central to human survival and well-being than water. According to recent reports from scientific organizations and multilateral agencies, the world's freshwater supplies are under increasing stress (Palaniappan and Gleick 2009). Of the total water resources on the planet—which measure about 1.4 billion cubic kilometers, a figure that is impossible for most people to contemplate—only about 2.5 percent consist of freshwater, and most of this precious resource is locked away in glaciers, groundwater, or vegetation, leaving only about 0.1 percent of the earth's water readily accessible for human use at any given time (Clarke and King 2004). Anthropogenic stresses such as population growth, urbanization, industrial pollution, rising consumption patterns, and inequitable distribution policies result in what has been termed the "global water crisis," which relates both to the quantity and quality of available freshwater (Cain and Gleick 2005; Gleick 2003). This crisis presents a particular challenge to economic development and human well-being in the developing world and constitutes one of the most pressing social and environmental problems of our time.

The problem stems from distribution as much as from simple supply and demand. The concept of "peak oil" has been used since the 1950s to describe the rate of oil production in relation to consumption levels. The pattern follows a bell-shaped curve, with time on the horizontal axis and annual production on the vertical axis: as demand rises, discovery and exploitation of additional sources rise to meet it. But this process causes costs to escalate, which makes further exploration prohibitively expensive, causing the rate of production to fall. The world has reached "peak oil" when at least half of the known stock of petroleum has been depleted and the rate of production begins to fall, driving prices ever higher.[7]

Water-resource experts have posed the question of whether water consumption follows a similar bell-shaped curve (Palaniappan and Gleick 2009). Are we facing an era of "peak water"? Like oil, water is essentially a fungible commodity: whether it is running through surface channels, seeping through bedrock, or coursing through vegetation as "green water," as long as two hydrogen molecules are bonded with one oxygen molecule, it is water. Unlike oil, however, the earth will never "run out" of freshwater because the hydrologic cycle ensures a continuous supply. But on a local or regional scale, people can and often do run out of freshwater. The North China Plain aquifer, which sustains a population of more than 100 million in and around the megacities of Beijing and Tianjin, has been tapped and used at a rate that far exceeds its natural recharge rate.[8] This trend is unfortunately true for surface water, too: some of the world's great rivers—the Colorado, the Rio Grande, the Nile, and the Yellow—are reduced to little more than a trickle by the time they reach the sea, overtaxed by agricultural irrigation, industrial use, and residential consumption.

The concern about the earth's water supply, therefore, is not simply about the consumption of a scarce resource. Rather, it is a concern about the distribution of that resource at the right times and in the right places to meet human demand without undermining the other important ecological functions that water fulfills. After all, water may be a fungible commodity, but it is not a substitutable one. We may be able to shift our dependence from oil to natural gas or some other fuel source, but in the portfolio of human needs nothing can take the place of water. And the human uses of water must always be balanced against the many other ecological services that water provides.

If we view water as a resource that is currently being used at or near its peak, the next logical question is how to manage it sustainably into the future. Peter Gleick (2003) suggests that various alternative management scenarios take either the "hard path" or the "soft path." The hard path comprises a one-size-fits-all model of large-scale hydrodevelopment consisting of centrally planned infrastructure projects that provide a reliable water supply, a source of power, and employment opportunities. The Tennessee Valley Authority, created in the United States during the Great Depression to enhance irrigation capacity and generate hydropower for a half-dozen states, is the epitome of such an approach, but it is not unique. The hard path has been the dominant paradigm in water-resource development around the world for about a century, promoted by government agencies and international financial institutions such as the World Bank.

The "soft path," by contrast, consists of improvements in the productivity and efficiency of water use, the application of economic principles to encourage efficient and equitable use, the development of new technologies of water management and distribution, and the engagement and participation of water users, communities, and stakeholders in the decision-making process. Practitioners of the soft-path approach seek to match the scale and character of water-development projects with the user's needs, to involve individuals and communities in important decisions about how water is used, and to balance environmental and human needs (Palaniappan and Gleick 2009:14). In the soft-path approach, bigger is not always better.

CRISIS AND OPPORTUNITY: THE STORY OF WATER IN CHINA

The anthropologist Eric Wolf has suggested that "the arrangements of a society become most visible when they are challenged by crisis" (1990:593). Such crises force people and institutions alike to make decisions about what is most important to them. In the introduction to an edited volume on the cultural dimensions of water resources around the world, Barbara Rose Johnston and John Donahue point out that the Chinese word for "crisis," *weiji*, is written with two characters: one that means danger or risk (*wei*) and another that connotes opportunity (*ji*) (1998:3). This is an apt metaphor for understanding the lengthy and complex story of water management in China, involving the acquisition, impounding, and appropriation of a natural resource to accomplish various national objectives.

Although water may be a "natural" resource, its allocation and use are inherently political, involving questions of power and justice. To this extent, water management is also the story of the evolution of human social institutions, especially markets and states, which allocate and control water, often by turning it into a commodity. The water crisis is therefore as much a social phenomenon as it is a biophysical one. As Donahue and Johnston note in the conclusion to their volume, "Crises involving water scarcity and water quality are as much a product of cultural values, social contexts, economic activities, and power relationships as they are a result of biophysical forces and conditions" (1998:345).

After decades of rapid economic growth, China currently faces a water crisis. In an unalloyed assessment of the issues, Gleick, who heads the Pacific Institute and produces a biennial report on global water assessment, concludes: "China's water resources are over-allocated, inefficiently used, and grossly polluted by human and industrial wastes, to the point that vast stretches of rivers are dead and dying, lakes are cesspools of waste, groundwater aquifers are over-pumped and unsustainably consumed, uncounted species of aquatic life have been driven to extinction, and direct adverse impacts on both human health and ecosystem health are widespread and growing" (2009:79).

The crisis relates to both water quantity and water quality. In terms of quantity, China's total renewable water resources and its land area are similar to those of the contiguous United States. However, because of its huge population, the annual per capita water availability in China is only 2,138 cubic meters per person as compared to 10,231 cubic meters per person in the United States (Gleick 2009:84). The distribution of water is also a constant problem: it is most readily available in the monsoon-prone southeast, while the heavily populated north suffers from chronic water shortages. The Yellow River, which has sustained Han Chinese settlements for thousands of years, dried up for the first time in 1972. In 1997, the river failed to reach the sea for 226 days, leaving a 700-kilometer stretch of riverbed dry (Molle et al. 2007:589).

It is a cruel irony that while the northern regions face chronic water shortages and the depletion of aquifers, many areas in the south must cope with the opposite problem: regular flood events during the monsoon season that cause loss of life, economic hardship, and ecological damage. The massive South–North Water-Transfer Project (Nan Shui Bei Diao Gongcheng), officially approved in 2002 and currently under way, is the epitome of hard-path solutions to this problem. The plan entails an interbasin transfer of massive amounts of water from the Yangtze to the Yellow, involving three routes to be constructed in different phases: the Eastern Route, consisting of a 1,200-kilometer-long canal, portions of which trace the path of the historic Grand Canal that connected Beijing with the port city of Hangzhou; the Central Route, which originates on the Han River, a major tributary of the Yangtze; and the Western Route, which would transfer water from the upper Yangtze basin, both from the main stem and from major tributaries such as the Tongtian, Yalong, and Dadu. Portions of the Eastern and Central Routes are currently under construction, but the feasibility of the Western Route, which traverses high-altitude, arid regions, remains in question.[9] Leaders are also trying to address water shortages through new policy mechanisms. The Three Red Lines (San Tiao Hong Xian) Policy, passed by the State Council in 2010, limits aggregate national water consumption to 700 billion cubic meters per year, mandates improvements in irrigation efficiency, and allocates 1.8 trillion yuan[10] during the Twelfth

Five-Year Plan period (2011–2015) for investment in water delivery infrastructure (State Council 2012).

Water-quality problems are no less pressing. Although China's environmental protection laws, policies, and institutions are well developed, enforcement is notoriously lacking. The MEP, which was elevated to full ministerial status in 2008, faces chronic shortages of funding and administrative weaknesses in comparison to agencies that drive the nation's economy, such as the NDRC. One large-scale example of the failure of environmental oversight and its consequences for water quality is the 2005 Songhua River benzene spill. An explosion at the Jilin Petrochemical Company, a subsidiary of the state-owned China National Petroleum Corporation, spilled nearly 100 tons of benzene, a known carcinogen, into the Songhua River in China's northeastern region of Manchuria. The spill necessitated the shutdown of the municipal water supply of Harbin, a city of more than 5 million people, and involved a government cover-up that resulted in the dismissal of the environmental protection minister.

Because water is one of the most fundamental elements of nature, it is tempting to view water problems—scarcity, quality, access, and distribution, among others—as equally natural. But a closer look reveals that water dilemmas, precisely because they involve such an elemental resource, are anything but natural. They are social, economic, and cultural. As Johnston and Donahue note, "Cultural notions, histories, economies, environmental conditions, and power relations all play a role in establishing differential resource relations, and this differential is a significant factor in ensuing conflicts and crises" (1998:5).

DAMS AS DEVELOPMENT

Like other kinds of water-engineering projects, dams have long been used by states and bureaucrats as tools for economic development. In fact, the case can be made that the successful rise of all large-scale civilizations depended to a great extent upon states' ability to mobilize resources and labor for allocating water where it was most needed. In ancient Mesopotamia, for example, the practice of sedentary farming and the eventual fluorescence of Mesopotamian culture across a vast area were made possible by extensive irrigation canal systems that required large-scale administrative coordination. The ancient Mesopotamian city of Mashkan-Shapir flourished because of a large network of canals connecting it to the Tigris River, allowing for the production of wheat and barley and the expansion of regional trade. After several decades, however, soil salinization—a common problem with irrigation projects in arid regions—caused the system and the city to collapse (Stone and Zimansky 2004). Similarly, gravity-operated water-distribution networks known as *aflaj* systems were commonly used in Oman and elsewhere in the Middle East to allocate groundwater for domestic use and agricultural irrigation. Many such systems are still in use today and rely on a combination of ancient engineering and traditional social management mechanisms largely outside the regulation of any state.[11]

Dams have occupied a prominent position in states' water-management portfolios for thousands of years: the remains of the Sadd-el-Kafara earthen dam in Egypt are dated to 2,600 B.C.E, and Roman dams constructed of concrete and mortar are nearly as ancient (Schnitter 1994). By the 1600s, the Spanish conquistadors thought they were taking European knowledge and expertise with them to the New World; when they arrived, however, they found that the Aztec capital Tenochtitlan was in its own right already a hub of water-engineering projects, including dams and aqueducts. Dams provide people with the opportunity to harness water for a variety of uses, including irrigation, flood control, household and commercial consumption, recreation, and navigation. They also equip people with the capacity to heavily use—and sometimes seriously overexploit—water resources. Dams are now a ubiquitous part of the landscape in most parts of the world: a georeferenced database of all dams and reservoirs worldwide shows a dense distribution of dams in almost every habitable part of the world (Nilsson et al. 2005).

The impacts of this proliferation of dams on the world's rivers have been profound. Reservoir storage exceeds the annual discharge in at least six major river systems: the Manicougan, Colorado, Volta, Tigris–Euphrates, Mae Khlong, and Río Negro. Another fourteen major rivers are so heavily regulated that more than half of their annual flow is diverted for reservoir storage and use. Several of the world's great rivers, including the Colorado, the Nile, and the Yellow, no longer reach the sea year round (Nilsson et al. 2005).[12]

In modern times, the construction of dams of all sizes is an integral part of the story of China's rise to industrial might. In 1949, on the eve of the socialist revolution that brought the CCP to power, the nation was home to only a handful of hydropower stations and reservoirs. In 1950, Mao Zedong approved the Guanting Reservoir Project near Beijing, which was completed in 1954 as China's first large reservoir project in the modern era (Ma Jun 2004:133). Understanding the prolific growth of dams in China over the past several decades requires a basic grasp of the nation's transformation from a largely agrarian economy to a manufacturing powerhouse, a trend that has been supported in part by a massive effort to regulate and harness rivers throughout the country. Approximately 86,000 dams of varying sizes were built in China between 1949 and 1990. The vast majority of them were embankment dams of relatively modest scale constructed of earth or rock and used to control flow, prevent seasonal flooding, or meet irrigation needs in a nation whose population was still primarily agrarian. Beginning in the 1980s and 1990s, however, the focus shifted to large-scale construction of hydroelectric facilities to help meet the nation's growing energy demand. The most notable and high-profile project initiated during this period is the Three Gorges Dam on the middle reaches of the Yangtze River, to date the single most expensive engineering project in the world.

Hydroelectric dams create electricity by converting the potential energy of water stored in a reservoir into kinetic energy, then converting that kinetic energy into electrical energy. Although the technological and engineering aspects of dams are quite complex, the science is relatively straightforward. Falling water, or water under pressure, is used to spin a turbine, which is connected to a generator that changes the mechanical energy from the spinning turbine into electricity through a coiled copper wire surrounded by a magnet. This process, called electromagnetic induction, has been understood since the early nineteenth century. The potential energy of the water impounded behind a dam is a function of the difference between the surface elevation of the reservoir and the elevation of the river below the dam, a concept referred to as "hydraulic head." Rivers with steeper gradients and high volume thus offer the greatest potential for energy generation.

In his book *Rivers of Empire*, which focuses on the arid western regions of the United States, the historian Donald Worster uses the term *empire* to denote the processes through which central authorities extract resources from peripheral regions and allocate them where they best serve national goals. In the past, this generally meant simply water appropriation: allocating a precious resource where it was most needed to accomplish national goals. However, in the case of hydropower development, it is more useful to think of water not simply as a commodity in its own right but as a source of kinetic energy that is transportable and can be used to fuel economic development in high-demand areas located thousands of kilometers away.

With rapid economic development, increasing demand for electrical power, and governmental support for large public-works projects, dam construction throughout the world reached its peak in the latter half of the twentieth century. Current figures indicate that 50,000 large dams, which ICOLD defines as those greater than 15 meters in height or having a storage capacity greater than 3 million cubic meters, exist in the world today (Scudder 2005; WCD 2000). China is home to about half of them (ICOLD 1998).[13] Measured in installed capacity, the total potential electricity output of hydropower facilities in China is more than that of Brazil, the United States, and Canada combined (Hennig et al. 2013).

Decisions made within China's hydropower sector—by government agencies and hydropower companies—tend to be technocratic in nature, driven more by engineering and economics than by broad public engagement. Indeed, technocratic governance seems endemic in the higher echelons of political leadership in China. As of 2000, more than half of all top political posts in China were filled by party members with engineering or technical college degrees. Li Peng, for example, who held top political posts during the 1980s and 1990s, including terms of service as premier and chairman of the Standing Committee of the National People's Congress, trained in hydroelectric engineering at the Moscow Power Institute in the 1940s and served for several years as minister of electrical power (Yeh and Lewis 2004:454).

As we examine hydropower development in the Lancang and Nu River basins in subsequent chapters of the book, the missteps, unintended consequences, and even human tragedies of this technocratic approach will become apparent. However, it should be noted that China is not alone in its penchant for pursuing technological and engineering solutions with little regard for their social, cultural, and ecological implications. The American legacy of dam building has proven particularly instructive to dam proponents around the world. Chinese scientists and policy makers who favor large-scale hydropower development in the southwest region explicitly cite Hoover Dam, Grand Coulee Dam, and other American examples as inspirations, noting in particular how the growth of California and other Western states was made possible by the huge expansion in dam construction in the 1930s, 1940s, and 1950s (Wang, Yu, and Li 2008). These dams, many of which began as job-creation strategies under the Works Progress Administration of the Great Depression, made new water sources available for expanding municipalities such as Los Angeles and new sources of electricity available to meet growing industrial and consumer demands.[14] Overcoming the aridity of the West through water-engineering projects was fundamental to the building of the American Empire (Worster 1985) but would also come to serve as a model for developing countries who wished to pursue hydropower development on a grand scale.

The Columbia River of the Pacific Northwest, America's fourth largest by volume, is the largest drainage system into the Pacific Ocean and the site of the most intensive hydropower production in the country. A film on the Columbia River dams entitled *Hydro*, released by the Bonneville Power Administration, features a segue between workers finishing the dam and a 1950s housewife, somewhere in the suburbs of the northwestern United States, plugging in her toaster and making breakfast for her family. Such is the relationship between increased availability of electricity and human well-being, as depicted in these thinly veiled propaganda pieces. As a resident of the Pacific Northwest region myself, I rely on Columbia River dams to produce much of the electricity that I consume every day. In the contemporary Northwest, environmental conservation, especially of Pacific salmon species, has become a topic of broad social and political concern, but in the heyday of dam construction during the mid–twentieth century the ecological costs of dams went largely unexamined. The great folk singer Woody Guthrie was commissioned by the Bonneville Power Administration to write an anthem celebrating dam construction on the Columbia River. Entitled "Roll on, Columbia, Roll On," it extols the virtues of early explorers such as Lewis and Clark, heralds the dams as symbols of national pride and progress, and points out how hydroelectricity made modern life in this region possible, ending each chorus with the refrain, "your power is turning our darkness to dawn."

However, looking back from the vantage point of today, we can see clearly a range of social and ecological costs that were not adequately considered when these dams were built. For example, Celilo Falls, a site of Native American habitation from the earliest days of human occupation of North America (more than 12,000 years before the present) and a significant fishing ground and trading post, was inundated by the construction of the Dalles Dam in the 1950s. By the time the dam was constructed, local Native Americans had already been dispossessed of their resources and involuntarily moved onto reservations. The Grand Coulee Dam, completed in 1942, effectively cut off 1,700 kilometers of salmon-spawning habitat in the upper reaches of the Columbia basin. Many books have been written that eulogize the Columbia, mourning its loss as a site of ecological treasure, cultural memory, and even spiritual significance (e.g., Harden 1996).[15] And the story of damage to ecosystems and displacement of native communities was repeated wherever dams like this went up in the United States.

Yet policy decisions are underpinned by human values, which constantly evolve. America is a different country today than it was seventy years ago. In 2007, several colleagues and I hosted an international conference on dams in Washington State, in a beautiful location along the Columbia Gorge about forty miles east of Portland, Oregon. As part of the conference, we visited the Bonneville Dam, the lowest of fourteen dams along the main stem of the Columbia River. As we toured the facility and descended by elevator into the powerhouse, where we could feel the great turbines shaking as they sent power out to the electrical grid, one of the members of our party asked the guide about strategies for improving fish passage. Fisheries issues, particularly for the five species of anadromous Pacific salmon, whose life cycle requires a traverse from freshwater rivers to the ocean and back again, have become paramount concerns in the Northwest. In response, the guide proceeded to describe several design characteristics, including the shape of the turbines and the presence of a fish ladder, and several operation strategies that manage the flow in such a way as to reduce fish mortality.

Impressed by this answer, another conference participant asked about the effects of these strategies on reducing fish mortality. Did such measures work? The guide responded that their efforts had reduced fish mortality to about 10 percent. I recall thinking briefly about what a triumph of engineering it was that nine out of ten fish could safely pass through the structure, until I remembered that there are more than ten dams on the Columbia. Upstream on the tributaries in Washington, Idaho, and Canada, the best historical spawning grounds in the Columbia watershed,

Coho salmon have gone extinct, and other species are listed as threatened or endangered.

But as societal values shift over time, so do the policy decisions that stem from them. In a subsequent conversation, a senior scientist at the Bonneville Power Administration—a federal agency under the U.S. Department of Energy, with responsibility for operating the Columbia River dams and distributing the electricity—told me that his organization takes in about $3 billion in annual revenues and spends approximately one-third of this sum, about $1 billion, on fish conservation. At the conference in the Columbia Gorge, the majority of presentations by American scholars focused on the deleterious effects of dams on river ecosystems. Several presentations even covered the topic of dam removal as an ecological restoration strategy, using case studies from the Pacific Northwest where the restoration of anadromous fish stocks such as salmon and steelhead has become a regional priority.[16] Our Chinese colleagues at the conference, even those from academic institutions and NGOs, were nonplussed: How could government agencies charged with economic development tear out dams to save fish?

WATER AND WORLDVIEWS

Water-management decisions are grounded in people's worldviews and the role that humans are perceived to play in the natural environment. Traditional Chinese views about nature—from Taoism, Confucianism, Buddhism, and even folk religion—tend to treat humans as integral parts of the natural realm. The history of environmental thought in China can be read largely as the record of contact, contrast, and amalgamation between these various streams of thought. Taoism, for example, emphasizes harmony and continuity between humans and the natural world they inhabit. The concept of *wuwei* (literally "inaction" or "not doing") describes the ideal relationship between humans and nature as one in which humans refrain from taking a dominating role and choose instead simply to exist in homeostatic balance with their surroundings (LaFargue 2001).

Confucianism, for its part, focuses on social order and a worldview that has been described as "anthropocosmic," which means that humans should maintain a delicate balance with the forces of nature. This relationship can be thought of as a triad consisting of heaven (*tian*), earth (*di*), and humans (*ren*) (Tu 1998). In contrast to Taoism, however, Confucian writings emphasize the pivotal role played by humans in the built environment: ordering the landscape, transforming it, and using it for economic purposes, especially cultivation. Where Confucian texts refer to resource management at all, they almost always prominently feature written characters with the radical denoting "agricultural fields" (*tian*) (Sommer 2012).

In the Buddhist tradition, the paramount value is the idea that humans should not harm other sentient beings. Grounded in the concept of karma, the "no harm" idea provides a basis for treating the environment with respect (Tucker and Ryuken Williams 1998). There is ample precedent within the main Chinese philosophical traditions for support of something akin to the "soft-path" approach to water-resource management. Chapter 25 of the enigmatic *Tao Te Ching*, for example, reminds the reader that "man follows the earth. Earth follows heaven. Heaven follows the Tao. The Tao follows what is natural" (Lao Tzu 1997). Confucianism sees humans and nature as intimately connected, embedded in historical fields of praxis and political relationships of hierarchy (Tu 1998).

The influence of this philosophical tradition on Chinese environmental practice is not hard to find. Less than 50 kilometers northwest of Chengdu, the provincial capital of Sichuan, is Dujiangyan, the great water-engineering project constructed more than 2,200 years ago. Its ingenious design was self-consciously Taoist, and Taoist temples still dot the banks of the river today. The Qin Kingdom (778–207 B.C.E.), a feudal state that emerged victorious from the Warring States Period and unified all of China under its rule in 221 B.C.E., was the administrative power in the region during the construction of Dujiangyan. Li Bing, an administrator during the waning years of the Warring States Period, initiated construction to put an end to the seasonal floods that made life along the Min River so precarious. Under his command, thousands of workers constructed the project, which consisted of two levees for flood prevention and one channel carved through solid bedrock that could divert part of the river's flow into a series of canals, bringing reliable irrigation to several thousand square kilometers of fertile farmland on the Chengdu Plain. This single feat is arguably what allowed the Qin Kingdom to surpass all other feudal states in terms of agricultural output and ultimately secure its place as the unifying force of the country. Heralded as an engineering marvel, Dujiangyan is still used effectively today for flood control and irrigation in the Min River basin.[17]

In modern times, however, these philosophical precedents exist in tension with a resource-management paradigm that is decidedly technocratic. There are currently hundreds of large dams under construction in China. The Three Gorges Project, the South–North Water-Transfer Project, and other similarly ambitious feats of engineering show China's propensity to frame resource problems as technocratic in nature, continuing a long and storied legacy of what might be called the "dictatorship of engineers." Reflecting on the Three Gorges Dam, the largest single site of electricity generation anywhere in the world, the environmental historian Donald Worster calls it "nothing new" (2011:5). Of course, to the 1.3 million people displaced by the reservoir, which extends hundreds of kilometers upstream, the dam is indeed something new. But Worster's point is well taken: this technocratic drive to harness the power of nature in the service of human needs constitutes more of a continuity with the past than a radical break from it. Sun Yatsen, father of modern China and first president of the Republic of China, had a vision in the 1920s of 100,000 miles of highway crisscrossing the nation and a hydropower dam spanning the middle reaches of the Yangtze River. The U.S. Bureau of Reclamation, in collaboration with the Nationalist government, helped conduct some of the feasibility and engineering studies for the project in the 1940s.

The capacity to control and regulate the nation's great rivers—for flood protection, for irrigation, for draught mitigation, and for navigability—has long held particular cultural significance within the collective memory of China's citizenry and long been a hallmark of dynastic strength. The rivers have not always cooperated. The Yellow River, which meanders from the arid interior to the Pacific carrying the huge load of loess sediment that gave it its name, was one of the earliest cradles of Han civilization. During a recent trip to the Chinese National Museum in Beijing, I viewed a Ming Dynasty painting that extolled the virtues of Li Xing, who oversaw a series of major projects to shift the channel of the Yellow River following a devastating flood in 1494 C.E. that killed thousands along the river's banks. Centuries later, Mao Zedong's famous historical couplet "When a great sage emerges, the Yellow River will run clear" [*Shang ren chu, huang he qing*]" referred to Mao himself, and the construction of the Sanmen Xia Dam seemed a natural exercise of this endowment (Shapiro 2000). In the end, large infrastructure projects such as dams are always about more than resource allocation or economic feasibility; they are often intended as grand public statements about the power of a state and its people to control nature. As Patrick McCully, director of the NGO International Rivers, has suggested, "They are concrete, rock and earth expressions of the dominant ideology of the technological age; icons of economic development and scientific progress to match nuclear bombs or motor cars" (2001:2).

NINE DRAGONS, SEVEN COMMITTEES, FIVE GIANTS

If hydropower development is a form of statemaking, sometimes it entails a recalibration of the relationship between state, society, and market. In contemporary China, harnessing the hydroelectric potential of major rivers and distributing the power on the electrical grid involve a mosaic of state agencies, including the State Council, the MWR, and the NDRC—the latter of which functioned as the main economic planning entity during the socialist era and now plays a major coordinating role. These administrative organs make rules, regulations, and policies on hydropower development. Depending on the specific details of a given hydropower-development plan, other agencies may play a role, including the MEP, the Ministry of Construction, the Ministry of Agriculture, the State Forest Bureau, China Guodian Corporation, the Ministry of Communication, and the Ministry of Health. When international rivers are involved, the Ministry of Foreign Affairs may also play a major role in decision making (Feng and Magee 2009).

Hydropower development is also a key part of China's "Develop the West" strategy (Xibu Da Kaifa), enshrined in the Tenth and Eleventh Five-Year Plans for economic development (2001–2005, 2006–2010), the goal of which is to narrow the economic and social disparities between the prosperous east coast and the relatively impoverished western regions, including Yunnan. The hydropower resources of the southwest region are vast; most of the nation's great rivers—the Yellow, the Yangtze, and the Mekong, among others—have their headwaters in this expansive, arid interior region of southwestern China. It is precisely these regions, moreover, that have been largely left behind in China's rush to develop its east coast, leading political leaders and hydropower officials to argue that dams can contribute both to the nation's energy supply and to jobs, revenue, and improved infrastructure in Yunnan.

These large-scale policy initiatives belie a tension between central-government control and economic liberalization and privatization, which have been the hallmarks of the reform era. Since the passage of the Water Law in 2002 (Chinese National People's Congress 2002b), China has been pushing for integrated river-basin management (IRBM, *liuyu zonghe guanli*), which the Global Water Partnership defines as "a process which promotes the coordinated development and management of water, land and related resources in order to maximize the resultant economic and social welfare in an equitable manner without compromising the sustainability of vital ecosystems" (UNEP 2008:5).

IRBM effectively broadens the set of stakeholders in water management and ensures that water-resource issues are considered in the economic-development plans put forward by other key state agencies, such as the NDRC. An approach that requires coordinated political and economic planning in order to achieve diverse objectives such as development, conservation, and social welfare, IRBM has been gaining momentum around the world over the past two decades. Its goal is to create institutions and processes that help overcome the fragmentation of water use for different purposes, bringing users into a common framework for allocating and managing water in a more coordinated way (Molle et al. 2007:587–588).

In practice, IRBM can prove "difficult to translate into operational terms" (Watson 2004:244). Chinese leaders have discovered that if they are to truly implement IRBM, they will need new management entities with a broader vision than any single agency can provide. Seven River-Basin Management Commissions (Shuili Weiyuanhui, most often translated as "Water Conservancy Commissions") were created in the 1950s. These commissions share oversight duties with provincial-level government agencies in the management of the nation's largest river basins: the Songhua, Liao, Huai, Hai, Yellow, Yangtze, and Pearl. They have regulatory control over water in major mainstream rivers, but not in tributaries. They manage water quantity, but not quality, which is the MEP's domain (Turner 2005). The River-Basin Management Commissions assumed new importance after the passage of the Water Law in 2002, which codified IRBM into national law, and they are now situated directly under the MWR, where they are responsible for the coordination of IRBM (Boekhorst et al. 2010).

In the spring of 2012, I visited the China Institute of Water Resources and Hydropower Research, which occupies a series of twelve-story steel-and-glass buildings on several city blocks on Beijing's west side. A row of black Audi sedans sat out front, their chauffeurs waiting. This is the epicenter of expertise, financing, and planning for water resources in China: the China Three Gorges Corporation is next door, the project offices for the South–North Water-Transfer Project are around the corner, the Sinohydro corporate headquarters is a block away, and the MWR is down the street.

In an interview, Dr. Yin, a senior engineer at the Institute of Water Resources and Hydropower Research, referred to this fragmentation of authority in the water sector as "seven committees, nine dragons." In ancient China, where the agrarian tradition and the Confucian ethical worldview favored large families with many male offspring, the saying about a dragon begetting nine sons (*long sheng jiu zi*) denoted a situation of extremely good fortune. But it was good fortune of a sort that could cause a host of practical problems: When it came time to divide the estate, for example, which son would inherit which parcel of land? When the parents became old and their health failed, which son would shoulder the responsibility of caring for them? It is a situation in which no single entity assumes ultimate responsibility.

This provides an apt metaphor, Dr. Yin suggested, for understanding the fragmented nature of multiagency management that now predominates in China's water sector. As with many sectors of the Chinese economy during the reform era, the political economy of hydropower development has been shaped by economic liberalization. In 1996, the National People's Congress passed the Electric Power Law (Chinese National People's Congress 1996), which required the power industry to begin disentangling electricity producers and distributors in an effort to create a nationwide market for electricity. In 2002, the State Electric Power Corporation, the key state-owned enterprise responsible for both power generation and distribution, was dissolved. Its assets and responsibilities were distributed among two groups of newly reconfigured state-owned enterprises: those responsible for electricity generation and those responsible for electricity transmission and distribution, the largest of which is the State Grid Corporation, which includes various regional subsidiaries. The group responsible for power generation comprises five state-owned enterprises, often called the "Five Energy Giants" (Wu Da Fadian Jutou), which hold diverse energy portfolios, including coal, wind, solar, and hydropower (see figure 2.2). Sinohydro, a major state-owned hydropower engineering and construction company, provides consulting services in research and development, design of equipment, and facility construction.

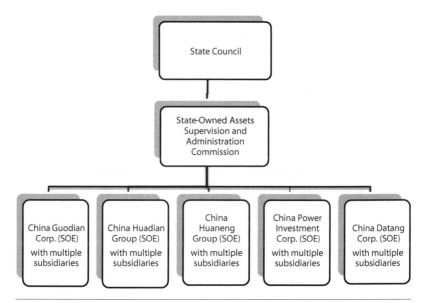

FIGURE 2.2 Current institutional structure of hydropower generation in China. SOE = state-owned enterprise.

Two of the Five Energy Giants are particularly important to the story of water management and hydropower production in Yunnan. China Huadian Group holds a state-granted monopoly on the right to develop dam projects on the Nu River through its subsidiary Yunnan Huadian Nujiang Hydropower Development Company. Meanwhile, China Huaneng Group holds the development rights on the Lancang River through its subsidiary Yunnan Huaneng Lancang River Hydropower Company. It views hydropower as a key strategy for what it terms "clean development [*qingjie fazhan*]" or "green development [*lüse fazhan*]" (Huaneng 2010b), and company documents list carbon-emissions reductions and climate-change mitigation among the many benefits of hydropower (Huaneng 2010a).

Although both Huaneng and Huadian remain state-owned enterprises, both are Fortune 500 companies with major subsidiaries that are publicly traded on the Hong Kong, Shanghai, and New York Stock Exchanges, and both enjoy close ties to the provincial government in Yunnan (Magee and McDonald 2009; Magee 2006).[18] In 2003, Huadian announced the formation of the construction entity Yunnan Huadian Hydropower Development Company, with registered startup capital of 200 million yuan (approximately U.S.$24 million). At this time, the shares were split between China Huadian Group (51 percent), Yunnan Development Investment Company (20 percent), Yunnan Electricity Group's Hydropower Construction Company (19 percent), and the Yunnan Nu River Electricity Group (10 percent) (McDonald 2007; Dore and Yu 2004).

This is a common trend in China's state-owned enterprise sector, which views the international market as a remedy for the chronic capital shortages it faces. The State Council, through a body called the State-Owned Assets Supervision and Administration Commission, exercises control over each of the state-owned enterprises, which assures that corporate governance is kept in line with the state agenda, an exceptionally close and complementary relationship between government agencies and key corporations (Jia and Tomasic 2009). Moreover, because the ownership structure of these corporations ultimately traces back to the State Council, lower-level agencies—including the MEP, whose task is to assess the environmental impacts of large-scale projects such as dams—find it difficult to regulate them effectively.

Like large companies everywhere, the Five Energy Giants constantly face the imperative to generate short-term shareholder value. For the major publicly traded subsidiary shareholder corporations, two types of stocks are traded in the market: "A shares," which are priced in Chinese yuan and limited to domestic investors, and "B shares," which are priced in U.S. dollars and open to both domestic and foreign investors (China Securities Regulatory Commission 2011). The State-Owned Assets Supervision and Administration Commission holds a controlling proportion of A shares, which ensures that the operations of the Five Giants remain in line with the state agenda.

This arrangement satisfies complementary interests between the Chinese government and foreign investors. From the government's perspective, allowing foreign investment ensures a steady supply of capital for key industries, many of which were state-owned enterprises that had long faced economic hardship. Statemaking sometimes requires a partial retreat of the state. For foreign investors, meanwhile, buying a stake in the Five Energy Giants is tantamount to betting on the continued expansion of the Chinese economy, given that these corporations hold diverse assets throughout the energy sector, from coal-fired power plants to solar, wind, and hydropower facilities. This investment is widely seen as a fairly safe bet because the expansion of energy consumption generally tracks the rate of annual GDP growth, which has hovered in or near double digits for more than three decades.

This complex network of public–private relationships is illustrative of the culture of economic liberalism that predominates, somewhat ironically, within the CCP. The rights to develop water-management infrastructure have been turned over to corporate interests, albeit ones in which the government holds a controlling share, while the overall logic and priorities of water management remain under the direction of the central government. The NDRC, the State Council, and other government agencies have the ability to approve particular projects and block others or to provide tax incentives to projects that further national economic goals.

SENDING ELECTRICITY FROM WEST TO EAST

In the case of Yunnan's rivers, hydropower production is not geared toward local consumption but rather toward supplying the eastern cities that now serve as the world's manufacturing hub, a fact that presents some unique political and geographical challenges. Although China's coastal regions have inherent geographical advantages in their position and even their historical links to the global economy, their rise to become the focal point of the modern Chinese economy was not inevitable; it was driven by explicit policy choices. Beginning in the late 1970s, as China began to engage with the global economy after decades of self-reliance under the socialist system, one of Deng Xiaoping and his political allies' first priorities was to establish Special Economic Zones along the east coast that would serve as manufacturing centers, magnets for foreign direct investment, and the cornerstones of a new, outward-looking economic model driven by exports. The most dramatic example of the effects of global capital on eastern China is the city of Shenzhen. In 1978, before its designation as a Special Economic Zone, Shenzhen was a medium-size city with a population of about 300,000 and an economy centered on fishing and agriculture. Today, its population, drawn from all corners of China and the world, numbers nearly 5 million, and its bustling economy is driven by manufacturing, investment banking, and international trade.

But the hydropower resources needed to fuel this industrial boom are not located in the floodplains and river deltas of the east; Shenzhen's factories and many thousands like them are fueled in part by the electricity generated on Yunnan's rivers, more than 1,500 kilometers to the west. Indeed, most of the nation's great rivers—the Yellow, the Yangtze, and the Mekong, among others—have their headwaters in the vast, arid interior regions of western China, far from the coast. These interior regions, moreover, have been largely left behind in the rush to develop coastal areas. In the "one country with four worlds" model, they represent the fourth world: vast regions of poverty in which people make a living principally by farming, wage labor (*dagong*), or out-migration. The central government has long recognized the social and political risks of such inequality and has begun taking steps toward addressing it with the "Great Western Opening" (Xibu Da Kaifa) campaign, which funnels investment into the western regions. Begun in the late 1990s, the campaign has become a major strategy for economic development in the West by harnessing the region's abundant natural resources, including oil, natural gas, minerals, and water resources (Lai 2002).[19]

Dams figure prominently in this overall strategy. They represent the convergence of central policy with topographical and geographical features that make the west most suitable for hydropower development. Whereas most of the dams in China have historically been embankment dams that store water for irrigation, located near population centers in the eastern and southeastern regions, this prevailing tendency is rapidly changing. A recent geospatial study of the distribution of dams in China over the past fifty years shows a clear trend: as hydropower surpasses irrigation as the principal policy objective in water-resource planning, dam construction is creeping inexorably westward in order to take advantage of the steep river gradients in western and southwestern China (Foster-Moore 2011:39).

The vast majority of electrical power from these rivers will be sent eastward to coastal cities in Guangdong Province, such as Guangzhou and Shenzhen, where industrial, commercial, and residential demands are high. This "Send Western Electricity East" (Xi Dian Dong Song) policy, initiated during the Tenth Five-Year Plan (2000–2005), is being facilitated by high-voltage direct-current transmission lines, cutting-edge technology that Chinese engineers have helped to develop and that allows electricity to travel long distances with minimal losses. As I show in subsequent chapters, there is considerable debate among scholars and policy makers about whether the Send Western Electricity East policy represents an economic boon to communities in the western regions or merely a perpetuation of existing regional inequalities, a form of "internal colonization" that promotes coastal development at the expense of historically marginalized areas (D. Goodman 2004). The Lancang and Nu Rivers, therefore, represent two crucial threads in the story of developing the hydropower capacity of the west to continue economic expansion in the east.

3 |

THE LANCANG RIVER

Coping with Resettlement and Agricultural Change

AS I noted in chapter 1, hydropower development in the Lancang basin is relatively far along, particularly in comparison with other nearby watersheds such as the Nu. Four dams have been completed—including Xiaowan, a massive concrete arch dam that, at nearly 300 meters high, is currently the world's tallest dam structure—and one is nearing completion. Taken together, these projects represent the development of about one-quarter of the basin's total hydropower potential. Tens of thousands of villagers have been resettled by the dam projects to date, and thousands more will be resettled in the years to come.

Displacement and resettlement, particularly by force, coercion, or government order, are the most controversial and politically sensitive social consequences of hydropower development. Understanding how the social consequences of displacement unfold over time is a task replete with methodological and political challenges, but crucial to formulating policy that accounts for the full range of costs and benefits associated with dams. Villagers resettled for dam projects in the Lancang basin, some more than twenty years ago and some much more recently, have faced the loss of farmland; a shift in income-generating activities away from agriculture and toward self-employment and wage labor; a disruption in the social networks of trust and reciprocity with other villagers that provide economic and social support; and a fundamental alteration in the cultural ties between people and place.

ECOLOGICAL AND CULTURAL DIVERSITY ON THE LANCANG

The Mekong is a transboundary river whose main stem and extensive network of tributaries support the livelihoods of tens of millions of people in China, Myanmar, Laos, Thailand, Cambodia, and Vietnam. Many sections of the river were mapped extensively beginning in the mid–nineteenth century, but other sections—in particular the upper Lancang in Yunnan—remained a mystery that inspired exploration on a global scale. A small group of French Jesuits stationed in Vietnam undertook an overland trek along the Mekong between 1866 and 1873, which would become known as the Mekong River Expedition (Osborne [1975] 1999). Their goal was to assess the feasibility of river navigation between Southeast Asia and mainland China. Although they failed to provide much meaningful information on the Mekong, they succeeded in establishing the fact that the Red River of central Yunnan could serve as a navigable channel for shipping between China and Vietnam. A Jesuit missionary named Doudart de Lagrée, one of the expedition's original leaders, serves as an illustration of the hardships this group endured as they traversed inhospitable terrain and interacted with people from a variety of Yunnan's ethnic groups who questioned their intentions. Through journal accounts published much later, we learn that Lagrée died an agonizing death, probably from amoebic dysentery, which caused a liver abscess. One of the expedition's physicians tried to drain the abscess in a crude surgical procedure under egregious sanitary conditions, which probably resulted in infection and expedited Lagrée's demise (Osborne 1975).

The story of the Mekong River Expedition serves to underscore the cultural and historical importance of the Mekong, both to the people who have lived along its banks and to the opportunistic individuals and institutions that would come to view it as a vital commercial route connecting much of Southeast Asia and southwest China. It was not until more than a century after the Mekong River Expedition that the precise source of the river, high on the Qinghai–Tibet Plateau, was pinpointed. In 1999, the Commission for the Integrated Survey of Natural Resources, a unit within the Chinese Academy of Sciences, determined it to be at a glacier at the foot of Guosongmucha Mountain at an altitude of 5,224 meters. By the time the Lancang reaches southern Yunnan's border with Laos, 2,200 river kilometers later, it has reached an elevation of lower than

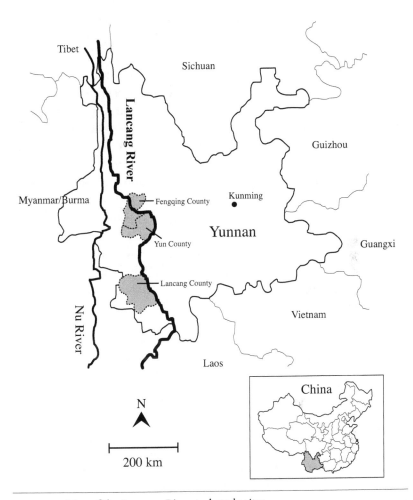

FIGURE 3.1 Map of the Lancang River and study sites.

1,000 meters, which highlights the extreme topographical relief and ecological diversity that characterize the basin. The river also traverses the
territory of at least a dozen minority nationalities, including Tibetan, Bai,
Dai, Yi, Lahu, Hani, and Bulang.

The anthropologist Guo Jiaji has suggested that we might think of the
myriad people living in the Lancang basin as distinct "ecocultures [*shengtai wenhua*]," shaped by long-term adaptation to diverse environments

and using a wide set of biological resources for survival (2008:96–118). This is an apt view, given the diversity between highland and lowland people in the watershed, along the river's north–south axis. From the highlands of the Qinghai–Tibet Plateau—a region that was historically known as Kham, the southeastern extent of the Tibetan culture area—through the middle reaches that support pastoral and agricultural livelihoods to the broad valleys downstream where villagers produce tea, tobacco, and rubber, the Lancang is the lifeblood of a diverse assemblage of cultural activities, livelihood strategies, and economies.

In addition to its economic and commercial allure, the northwest corner of Yunnan has long been an object of aesthetic wonder and even spiritual fascination for Western explorers and missionaries. Zhongdian County, in the northwest corner of Yunnan, was renamed Shangri-La (Xianggelila in Chinese) in 2002 as part of a government program to promote tourism development. The name derives from the famous James Hilton novel of the 1930s, *Lost Horizon*, in which a British diplomatic services officer, stranded after an airplane crash, finds inner peace and eternal youth in a Buddhist lamasery located on the Tibetan frontier. Among Shangri-La aficionados, some controversy remains about the novel's precise setting and about what location therefore constitutes the "real" Shangri-La, which may in fact be entirely fictional. Nevertheless, the notion of a unique and strikingly beautiful landscape inhabited by people from exotic and poorly understood cultural groups has become a powerful mythology, attracting tourism revenue from foreign and Chinese travelers alike (Litzinger 2004; Hillman 2003).

Scientific discovery also played a prominent role in the West's understanding of this corner of Yunnan. Northwest Yunnan and the Tibetan frontier were the stomping grounds of Francis Kingdon Ward, an English botanist and explorer who mounted many expeditions in the region over a span of nearly fifty years beginning in the early 1900s. He came in search of floral specimens and seeds to carry with him back to England, and dozens of local plant species still bear his name. His writings provide some of the earliest accounts of Western exploration in northwest Yunnan. The Hengduan Mountains form a series of high divides between the watersheds of the Jinsha, Lancang, and Nu Rivers. A keen observer, Ward noted the significant differences in precipitation, flora, and fauna between these watersheds. In a book entitled *Mystery Rivers of Tibet*, Ward recounts his journey of 1913 overland from Zhongdian and westward to the upper reaches of the Lancang: "We camped that night at 13,000 feet amidst patches of melting snow and frozen-looking Rhododendrons, where tortured larch trees wrestled to the death with the inhospitable climate. On June 2nd we crossed a pass and reached the rolling plateau at the top of the Yangtze–Mekong divide. The summit of the range is conspicuously glaciated. It was a dreadful day, swift showers sweeping up the valley on the wings of the wind; eastward the snowy peaks of the Pai-ma-shan range were muffled in cloud" (1923:32).

Despite its location on the periphery of the Chinese Empire, or perhaps because of it, Yunnan has long served as a conduit for economic and cultural exchange between China, Southeast Asia, and Central Asia. The ancient Tea-Horse Road (Chama Gudao), an important trading route since the days of the Han Dynasty (206 B.C.E.–220 C.E.), traced the banks of the Lancang for part of its course before arcing westward into Tibet and Central Asia. Tea porters, most of whom were Tibetan or Naxi, carried heavy loads of compressed tea cakes over hundreds of kilometers of inhospitable terrain, a way of life that persisted into the early years of the twentieth century. They traded their tea for horses, thus satisfying both sides: the Tibetans, who could mix the tea with barley flour and yak butter to make *tsampa*, and the Chinese, who quickly became accustomed to using horses in their daily activities for agricultural traction and transportation.

Beyond the imaginings of Western scientists and explorers, of course, the people who live in the Lancang basin have their own origin stories that provide a basis for attachment to place. In the Tibetan areas along the northern extent of the river, villagers herd yaks and *dzos*, a hybrid between yaks and domestic cattle, using the milk and meat from these animals for subsistence and market exchange. Some households cultivate potatoes and barley, some of the only crops that will grow reliably at elevations higher than 3,000 meters, and collect nontimber forest products, including wild herbs, mushrooms, and, most famously, caterpillar fungus (known in Chinese as *dongchong xiacao*, literally "winter insect, summer grass"). The fungus, which includes various species belonging to the genus *Cordyceps*, has a very peculiar life cycle: fungal spores invade the tissue of a caterpillar, where they lie dormant over the winter; eventually, in late spring or early summer, the fungal fruiting body sprouts from the corpse of the worm, whose body has become food for the parasitic fungus. In traditional Tibetan medicine, *Cordyceps* was used to treat a variety of maladies, including asthma and high blood pressure; today, on the Chinese market, where it is prized for its aphrodisiacal properties, it can sell for as much as 500 yuan per gram, providing local Tibetan households with up to half of their annual cash income and supporting bustling markets and distribution networks (Galipeau 2012). Herbal medicine stores from Kunming to Beijing advertise the health benefits of this curious product, allocating prominent shelf space to glass jars full of *Cordyceps*.[1]

The anthropologist Lun Yin, who has conducted fieldwork in the region's Tibetan communities, has documented the cosmological connections between the people and the Lancang River:

> The Tibetan people describe the nature of their four main rivers in a series of important myths. The Nujiang River is the beautiful mother of all rivers, originating in Tibet and flowing south, and the Dulongjiang, Jinshajiang and Lancangjiang Rivers are her three sons. The Dulong-jiang is the bad-tempered eldest son, who brings severe disasters and diseases to local people. The god of Kawagebo Mountain punished such bad temper by changing the Dulongjiang's flow from the south to the west. The Jinsha River is the bold and uninhibited second son, bringing fortune and gold to the local people. The huge wave sounds made by this arrogant river enraged the god of Kawagebo, who changed the river flow from south to east. The last son is Lancangjiang, the Mekong River. He is both kind and peaceful, bringing abundant fruits and agricultural products to those who live along his banks. The stories of the rivers tell people what to expect from their natural environment and how they should behave towards these primary sources of life-giving water.[2]
>
> (2012:186)

The Lancang is indeed quite peaceful and tame in comparison to the other rivers in this mythic story. The upper reaches of the Lancang gorge provide a hospitable climate that supports paddy rice production in some places. Downstream, at the southern extent of Yunnan, the Lancang passes through Xishuangbanna, a Dai Autonomous Prefecture known throughout China for its subtropical climate and floral diversity. The Dai traditionally maintained this agro-ecosystem through swidden practices, but much of the farmland has been converted over the past several decades to rubber, palm oil, and tea plantations (Guo et al. 2002).

LANCANG DAMS: PAST, PRESENT, AND FUTURE

The exploitation of the Lancang River basin for its hydroelectric power potential commenced on a modest scale nearly seventy years ago. The Tianshengqiao Hydropower Station, a small facility with an installed capacity of 400 kilowatts, was completed in 1946 on the Xi'er River, a tributary of the Lancang. Since the 1980s, the scale and pace of hydropower development have gained momentum on the Lancang's main stem, with a "lower cascade" of seven dams planned (see table 3.1).[3] Two of these dams, Xiaowan and Nuozhadu, are among the world's tallest arch dam structures, taller than the Three Gorges Dam, and create very large reservoirs, inundating vast tracts of land and forcing the resettlement of thousands of people (Magee 2006).

Manwan Hydropower Station, with an installed capacity of 1,500 megawatts, was completed in 1996. Dachaoshan Hydropower Station, with an installed capacity of 1,350 megawatts, was completed in 2003. The Xiaowan Hydropower Station, with an installed capacity of 4,200 megawatts, was completed in 2012. Jinghong Hydropower Station, with

TABLE 3.1 Design and Operation Specifications for the Seven Dams in the Lancang Lower Cascade

Dam Name	Height (m)	Installed Electrical Capacity (megawatts)	Reservoir Storage Capacity (million m3)	Estimated Population Displaced	Completion Date
Gongguoqiao	130	750	510	4,596	2015
Xiaowan	292	4,200	15,130	28,748	2012
Manwan	132	1,500	1,060	3,042	1995
Dachaoshan	121	1,350	880	5,200	2003
Nuozhadu	260	5,850	22,740	14,800	2017
Jinghong	260	1,750	1,230	1,700	2011
Ganlanba	61	155	72	58	2015
Total		15,555	41,622	58,144	

Note: Dams are listed from north to south.

Source: Magee 2011; He, Hu, and Feng 2007:152.

an installed capacity of 1,750 megawatts, became fully operational in 2011 (He and Chen 2002). Nuozhadu Hydropower Station, which will have the greatest hydropower capacity, at 5,850 megawatts, is under construction, with an expected commissioning date sometime after 2014. Taken together, these projects represent the development of 28 percent of the Lancang basin's estimated hydropower capacity (Liu 2012).

Along the upper sections of the Lancang, government agencies and hydropower corporations have made plans for an "upper cascade" of dams, ranging in number anywhere from five to eleven (Magee 2011). It remains controversial in part because of the excessive cost of building roads and other associated infrastructure in an extremely underdeveloped region. The news media often report stories about the cancellation of one or more upper-cascade dams, and it is unlikely that all eleven will be built. However, it appears that preparatory work is proceeding on at least two of the upper-cascade dam sites—Lidi and Wunonglong. In 2012, a journalist who had recently visited the sites for a magazine story showed me photographs on her tablet computer revealing hillside scars, road construction, and other signs of preparatory work at these two sites.

While the Lancang dams have attracted controversy in their own right, they have also raised concerns about a so-called domino effect in which downstream riparian countries, already affected by the altered flow regime of the Mekong, begin dam projects of their own in order to reap the benefits of electricity and revenue. After years of debate, the People's Democratic Republic of Laos announced in late 2012 that construction would begin on the Xayaburi Dam, the first of what will likely be nearly a dozen projects along the middle and lower reaches of the Mekong in the years to come (Ngo 2012). As I show in subsequent chapters, all of the countries in the Greater Mekong Subregion have devised plans for damming the river, and many rely on Chinese expertise and financing.

LEARNING FROM THE MANWAN DAM

Much of what we know about the effects of dams on communities in Yunnan comes from case studies of the Manwan Hydropower Station, which was completed in 1996 and has been the subject of fairly comprehensive studies over the past two decades. At 132 meters high and 482 meters across, it is actually one of the more modest projects on the river. Upon its completion, however, the Manwan reservoir extended 70 kilometers upstream, inundating more than 6,000 *mu* of farmland and 8,000 *mu* of woodlands across 114 villages in eight townships within Jingdong, Yun, Fengqing, and Nanjian Counties.[4]

The Manwan Hydropower Station was initiated as a joint project between the Yunnan provincial government and the MWR but has since been turned over to Yunnan Huaneng Lancang River Hydropower Company (known in English as Hydrolancang), a subsidiary of China Huaneng Group, one of the Five Energy Giants, which holds a state-granted monopoly on hydropower development rights in the Lancang basin (Magee 2006). A social and environmental impact study, sponsored by Oxfam Hong Kong, was conducted during the summer of 2000 and incorporated into a document entitled *Reasonable and Equitable Utilization of Water Resources and Water Environment Conservation in International Rivers in Southwest China*, a key component of China's Ninth Five-Year Plan for Science and Technology.[5] The product of collaboration between natural and social scientists, this study provides a baseline for understanding the social and ecological effects of the Manwan Dam.

The findings of the social impact assessment team, which included a number of Chinese social scientists from government institutions, were striking. Before construction began, government officials had estimated that 3,042 people, mostly from local farming households, would be displaced. However, the actual figure totaled more than 7,000. This is a common trend worldwide, where dam-resettlement programs often underestimate the number of potentially displaced people, which can be especially problematic when budgeting for compensation programs is based on the initial lower estimate.[6] Government officials coordinated the resettlement effort, with households classified into five types: those resettled in villages outside the reservoir region; those resettled in villages within the reservoir region; those resettled into nearby cities and towns; those who could remain in their villages with a new allocation of farmland; and those who could remain in their villages and did not require farmland reallocation. Some of the most productive paddy fields with intensive irrigation systems were inundated, resulting in decreased yields of rice and other staple grains, forcing local people to shift to the production of maize, sugar cane, and other dryland crops.

The per capita incomes of resettled households in the Manwan area quickly fell to less than half of the provincial average, which already represented some of the lowest income levels in all of China. Many residents turned to wage-earning jobs outside their home communities in construction, tourism, and related industries, sending remittances back to their families. Ironically, chronic electricity shortages continue to plague the area adjacent to the Manwan Dam, which is geared primarily toward sending electricity to the booming cities of Guangdong Province under the Send Western Electricity East policy. Although every resettled village has been connected to the power grid, local residents can purchase electricity supplied by the Manwan facility for RMB 0.9–1.5 per kilowatt-hour, which is several times more expensive than the electricity previously supplied by a micro–power station on a small tributary, a facility that was inundated by the reservoir and is now inoperable.

Dr. Guo Jiaji, of the Yunnan Academy of Social Sciences, led a longer-term study on resettlement for the Manwan Hydropower Station, with the goal of providing suggestions to government authorities on how their policies and practices could be improved. Faced with an inundation area that affected more than one hundred villages, the research team selected three representative villages—to which they gave the pseudonyms Dam Village, River Village, and Cliff Village—as in-depth case study sites. Guo's study provided, for the first time, the data necessary for a direct comparison between the objectives of the resettlement program and its actual outcomes.

The Yunnan provincial government, in cooperation with Huaneng Corporation, committed to invest 50 million yuan to address the social and economic problems associated with resettlement. These funds, according to government reports, would be allocated to accomplish an ambitious set of goals that included building or renovating 100 schools; building 100 village sanitation clinics; establishing 100 village cultural centers for recreation; subsidizing educational training programs for 1,000 rural teachers; subsidizing the tuition of 1,000 high school graduates to attend technical school; paying tuition for 1,000 elementary and middle-school students from designated "poverty households" (*pinkun hu*); subsidizing a potable-water-development project that would provide clean drinking water to approximately 10,000 people; subsidizing employment retraining for 10,000 displaced people; and subsidizing the enrollment of up to 100,000 people in China's New Rural Cooperative Medical Care System, which would increase villagers' access to local health-care clinics (Guo 2008:205).[7]

The general practice during resettlement was one in which the upper level of government would "practice oversight," and the local government would "seek truth from facts [*Dui shang bao gan, dui xia shi shi qiu shi*]," a Dengist slogan that is essentially pragmatic: do whatever is necessary to make things work, without regard for ideology (Guo 2008:203). This meant that the provincial government would set the guidelines for resettlement and compensation, but that the county and township governments would have some flexibility in carrying them out. In the end, the resettlement program fell far short of its ambitions. According to Guo's research, villagers with adequate social connections (*guanxi*) to political officials were able to secure jobs working in the hydroelectric facility or to pursue other economic opportunities, while most villagers either joined the throngs of migrant laborers in nearby towns or farmed marginal land plots on steep hillsides without irrigation. Displaced villagers faced long-term difficulties in securing what many rural people think of as a triad of basic needs: eating (*chi*), clothing (*chuan*, literally "wearing"), and living (*zhu*). With their land inundated, their income in decline, and the price of grain on the rise, many villagers who were accustomed to eating rice had to switch to eating corn; one villager noted, "We're not starving yet, but the food we eat leaves a void in our hearts. After eating this month, we're not sure what we're going to eat next month" (qtd. in Guo 2008:210).

In regards to clothing, people from nearby villages commented bitterly that residents of Dam Village could often be seen wearing fancy clothing. In response, one resident of Dam Village replied, "Our clothing includes leather items worth several hundred yuan. But it's all scavenged from the garbage bins of those who work in the hydroelectric plant" (qtd. in Guo 2008:210). Housing for resettled villagers was a chronic problem involving the confluence of ecological factors, the changing winds of political economy, and extremely bad luck. The initial feasibility studies for Manwan were conducted in the 1980s and concluded that the government would need to provide about 17.6 million yuan to compensate displaced migrants. However, the government failed to account for the skyrocketing changes in the economy brought on by Reform and Opening or for subsequent upward pressures on living costs. The cost of building materials such as wood and cement increased dramatically during the boom years of the 1990s, which made it difficult for villagers to build houses matching the size and quality of what they were accustomed to, even with housing subsidies provided by the government. To make matters worse, multiple years of monsoon flooding caused damage to several villages resettled on sloped land that had been cleared for housing developments; the houses were condemned as unsafe, and their occupants were forced to move a second time. Primary-school students from one village, whose flood-damaged school was condemned as unsafe, had to travel several kilometers to Manwan Township in order to attend school.

Studies such as Guo's that seek to make policy recommendations on resettlement issues, tend to approach the issue with caution and circumspection. The authors tend not to overtly criticize central policy or the overall objective of building hydroelectric dams but rather try to provide a detailed analysis of the outcomes of resettlement, make recommendations, and advocate for better participation by those most affected by resettlement. In subsequent chapters, I revisit the issue of whether policy makers have taken these lessons to heart and how resettlement policy can better address some of the problems outlined here.

CONTEMPORARY LIVELIHOODS ON THE LANCANG

In 2009, I traveled to the middle reaches of the Lancang with a group of American and Chinese researchers to conduct household surveys in Fengqing County (the site of Xiaowan Dam), Yun County (the site of Manwan and Dachaoshan Dams), and Lancang County (where Nuozhadu Dam was under construction at the time). We headed west from Kunming by bus on an expressway, about 270 kilometers to the town of Midu, then hired a smaller car to go another hour south to the county town of Nanjian, and then for another two hours went by local bus to Xiaowan Township, the location of the Xiaowan Dam and many of the resettlement communities associated with the project.

Near Nanjian, the wide valleys were planted in rice, corn, sweet potatoes, tobacco, and other crops. The rice stalks, a deep green hue in early summer, were nearing their full height. As we approached the Xiaowan Dam after a long day of travel, the terrain became progressively more rugged, and the bus's engine labored as we wound in and out of steep gorges, the slopes of which were terraced for farming. The serpentine river—brownish gray and laden with sediment—came into view, disappeared, and reappeared. Along the way, small-scale industries dotted the landscape: a gypsum quarry, a cement plant, a coal mine. For the final hour of the ride, the road traced the path of huge power-transmission lines, buttressed by steel scaffold towers, arcing their way eastward toward the urban manufacturing hubs in Guangdong Province.[8]

The rainy season had begun, and television and radio reports were warning of landslides in the area. Through sheets of heavy rain, we viewed the Xiaowan Dam, mostly complete at that time except for the installation of a few turbines, from a bridge a half-kilometer downstream. The massive structure—at 292 meters currently the world's tallest dam—could be seen through the dense air, wedged into a narrow section of the canyon. Several families were walking along the roadside on a bluff overlooking the dam, their group framed by hillside scars and a new access road that had been carved into the mountainside in the distance. Winding along the road several kilometers upstream from the dam was the reservoir, which at that point was just beginning to fill, raising the water level in several tributaries like a hand with swollen fingers.

Xiaowan Township has grown tremendously over the past decade, as engineers, construction workers, road pavers, and business entrepreneurs have made their way to the area. In addition, many farming households from outlying villages within the reservoir area have been moved into the township, where they live in newly constructed, multistory cement houses within what locals refer to as the "migrant village" (yimin cun). The township government offices, formerly located on the left bank of the Lancang, were recently moved to a new position on the right bank, atop a prominent hill above the township, where the national flag flew on a flagpole. A new housing development of two-story brick villas, containing beautiful courtyards with flowering bougainvillea and camellia shrubs, had been constructed nearby; its housing units were filled primarily with government officials and high-level hydropower company employees. Water buffalo grazed lazily in the fields and sometimes walked carelessly down the middle of the road, which they shared with black government sedans and a steady stream of yellow four-wheel-drive vehicles featuring the China Southern Power Grid Company logo above the company slogan: "Lighting Ten Thousand Homes, the Love of China Southern Power Grid Runs Deep" (Wan jia denghuo, nanwang qing shen). The township center was full of commercial activity and businesses, including restaurants, clothing shops, barber shops, laundry services, a karaoke bar, and even an Internet café, where young people played videogames late into the night.

FIGURE 3.2 A farming family's house in Fengqing County, with beans drying in the courtyard.

In the outlying villages—which were connected to one another by a series of winding footpaths traversing rice paddies, mixed-crop fields, and tea plantations—most houses retained the traditional construction features of rural Yunnan: rock foundations, thick adobe walls, and ceramic-tile roofs. Many houses had only dirt floors, but others had poured concrete. Village women cooked over wood fires, using firewood when available or scrap wood and crop residues such as corn cobs when supplies ran low. During one household interview with a middle-aged woman who had just returned from the mountainside wearing a goatskin pack containing a bundle of wild herbs to feed to her pigs, we looked around the house, which comprised a storage room with bags of rice and wheat; a fairly large kitchen; a main sitting room with a television set; a bedroom with several beds; and an outhouse toilet. Like most local villagers, she spoke a regional dialect that our Chinese colleagues, even those from Kunming, had difficulty understanding.

The courtyard of the house was filled with recently harvested beans set out to dry. A hindquarter of brined pork was suspended by a length

TABLE 3.2 Basic Characteristics of the Lancang River Study Sample

County	Household Count (Sample %)	Dam Sites (Status)	Ethnic Composition*	Literacy Rate	Households Displaced (% of Sample)†
Fengqing	199 (26.2%)	Xiaowan (completed 2010)	Han: 82% Yi: 16% Other: 2%	97%	62 (31%)
Yun	357 (47.0%)	Manwan (completed 1996); Dachaoshan (completed 2003)	Han: 78% Yi: 15% Bulang: 4% Other: 3%	92%	124 (35%)
Lancang	203 (26.7%)	Nuozwhadu (under construction)	Han: 24% Yi: 22% Lahu: 51% Other: 3%	96%	60 (30%)
Total Sample	759 (100%)			95%	246 (32%)

*Other ethnic groups represented in the sample include the Dai, Bai, Hui, Wa, and Yao.
†Percentages have been rounded to the nearest whole number.

of rope from the rafters to cure. A swallow's nest of mud and straw was nestled under the eaves of the house; beneath it, the household head and several of his kinsmen sat talking and smoking tobacco, which had been harvested and cured in the village, in bamboo pipes, their pant legs rolled to the knee against the summer heat and humidity.

Our survey team collected data from 759 households across Fengqing, Yun, and Lancang Counties (see table 3.2), typically interviewing either the male or female household head.[9] We asked a wide range of questions about demographic, social, cultural, and economic conditions; agricultural production; participation in village activities; and people's social networks. Within each county, the survey team tried to capture in the sample both resettled households and households that had not been resettled. Villagers in the study represented ten *minzu*, including Han, Yi, Lahu, and Bulang.

All three counties boast historically strong records of agricultural production. Most households produce rice as a staple crop for household consumption and for market sale, and many also produce corn or a winter wheat crop, plus a variety of commodity crops such as sugar cane, mangos, melons, chestnuts, and even rubber. With the rapid marketization of Chinese agriculture over the past several decades, many households in the Lancang basin grow at least some of what local villagers call "the big three" (*san da*): walnuts, tea, and tobacco. A government official in the Fengqing County Office of Financial Affairs pointed out that the more successful farming households tended to take advantage of "agricultural marketization" (*nongye shichanghua*) by forgoing rice production in order to focus on earning a cash income through the sale of the big three commodity crops. Walnuts in particular were fetching a very good price at the time: 12–13 yuan for one *jin* (half-kilogram). The local tea, a slightly different variety than Yunnan's famous Pu'er Tea, was selling for 7–8 yuan for one *jin*. Of course, tree crops such as walnuts and tea require a considerable upfront investment and entail the opportunity cost of waiting for several years as the trees mature enough to produce a crop.

AGRICULTURAL LIVELIHOODS IN TRANSITION

One major goal of the survey effort was to determine how resettlement affects agricultural livelihoods in the Lancang River basin. When a dam is built and a reservoir fills behind it, inundating people's homes and farmland, how do their landholdings, cropping strategies, and income change, and what does this change mean for their future? Answering these questions turns out to be more difficult than it first appears. Researchers seeking to understand the effects of dam-induced resettlement on communities face some serious methodological challenges. Ideally, one would collect longitudinal data from communities before and after resettlement, documenting how social and economic conditions vary from a known baseline. But the effects of resettlement typically unfold over a long time horizon, often many decades, making this approach impractical without an army of researchers and an unlimited budget. Having neither, our research team opted instead for a cross-sectional study approach that compared resettled communities with nearby communities that had

TABLE 3.3 Agricultural Livelihoods and Income in the Lancang River Study Communities

		Not Resettled ($n = 513$)	Resettled ($n = 246$)
Land Holdings (*mu*)	Paddy Land[†]	1.0	1.8
	Dry Land[†]	15.9	7.0
	Forest Land[†]	13.2	3.3
Staple Crop Sales (yuan)	Rice[†]	434	1,039
	Corn[†]	5,624	3,626
Commodity Crop Sales (yuan)	Walnuts[†]	613	4
	Tea	260	33
	Tobacco[†]	1,462	3,625
Household Income (yuan)	Agricultural Income	19,177	15,026
	Livestock Income	3,068	3,647
	Wage Labor[†]	5,181	14,105
	Self-Employment[†]	3,057	9,071
	Total Household Income[†]	30,565	41,850

Note: Independent-samples t-test ([†] significant at 0.01 level).

similar demographic characteristics but had not undergone resettlement. This cross-sectional approach did not allow us to directly measure diachronic change in agricultural livelihoods for any given household, but it did allow for a systematic examination of differences between households based on resettlement status.[10]

Of the households in the sample, about one-third (32.4 percent) had been resettled at the time of the survey. Table 3.3 compares resettled and nonresettled households based on their land holdings, crop sales, and income. In regards to land holdings, resettled households held more paddy land (+0.8 *mu*), less dry land (−8.9 *mu*), and far less forest land (−9.9 *mu*) than their counterparts who had not undergone resettlement.

This pattern of land access coincides with some dramatic differences in cropping strategies between the two groups. Resettled households produced and sold more rice, which is grown on paddy land, and less corn, which is a dryland crop typically reliant on intermittent seasonal rainfall or labor-intensive irrigation by hand. In terms of the "big three" commodity crops, the two groups also differed considerably: resettled households produced more tobacco than nonresettled households but lagged far behind in the production of tea and walnuts. This makes intuitive sense because tea and walnuts are tree crops that require a significant amount of upfront capital and a number of years in situ before the trees are mature enough to produce a crop.

Perhaps most significantly, the two groups show remarkable differences in their participation in off-farm labor. Resettled households were twice as likely to report having at least one member of the family participating in wage labor, often as a migrant worker in a town or city. For these families, remittances (14,105 yuan per year on average) had become a crucial income source. Resettled households also participated much more in entrepreneurial, self-employment activities, such as operating restaurants or retail shops, and earned three times more self-employment income than their nonresettled counterparts. As a result, total household income, which includes both cash sources and in-kind sources such as the value of agricultural products grown and consumed at home, was much higher for resettled households than for their counterparts who were unaffected by displacement.[11]

This finding contradicts much of what social scientists have learned in recent years about the consequences of development-induced displacement for local communities. The anthropologist Thayer Scudder (2005), for example, who has undertaken perhaps the most comprehensive and far-reaching review of the social impacts of large dams around the world, argues that there are very few cases in which dam-induced displacement resulted in improved livelihoods for local people. One key point of the controversy over the long-term socioeconomic effects of dams is whether dam construction and operation—the most labor-intensive stages in a project's life cycle—actually create jobs for local residents, an argument that government agencies and hydropower corporations routinely put forward in support of their agendas.

The picture in the Lancang basin appears to be mixed. In Yun County, among households reporting some income from wage labor, fewer than 4 percent had someone in the household who held a job connected to a hydropower facility. In Fengqing County, no households reported a member working in a job connected to hydropower. In Lancang County, however, 40 percent of households with a member working in wage labor said that someone in the household worked in a job related to hydropower, likely because the Nuozhadu Dam was still under construction at the time the survey was administered, providing more opportunities for unskilled work. If Lancang County follows the typical pattern of other dam projects in China and elsewhere, low-level job opportunities will likely disappear once the dam begins operation. I revisit this issue in chapter 6 with a closer examination of resettlement-compensation policies and their effects on households in the Lancang basin.

DISPLACEMENT AND SOCIAL NETWORKS

Because dams uproot communities, they have the potential to seriously disrupt social networks, the webs of interdependence that community members maintain with one another through relationships of trust and reciprocity. Social scientists have been interested in the study of social networks as an intellectual project going back at least to Emile Durkheim's work in the late nineteenth century. These networks provide a basis for communal identities, but they also serve more pragmatic and instrumental purposes, helping to distribute resources, disseminate information, and provide economic support in good times and bad.

Such networks reinforce a community's ability to adapt to changes or external stressors; they "provide a basis for social cohesion because they enable people to cooperate with one another—and not just with people they know directly—for mutual advantage" (Field 2003:12). This is particularly true in China, where dyadic social ties colloquially known as *guanxi* are a tremendously important means of securing a job, navigating the political system, and otherwise making one's way in society (M. Yang 1994). How might the resettlement process affect villagers' social networks? To address this question, the survey asked villagers about two aspects of their networks: their reciprocal relationships with others, including the sharing of financial resources and farm labor; and their subjective attitudes about community life in their village.

Table 3.4 shows a comparison of reciprocal behavior between resettled communities and nonresettled communities over a twelve-month period. I conducted basic statistical analyses, using independent-samples t-tests

TABLE 3.4 Comparison of Reciprocal Behavior Between Resettled and Nonresettled Communities

Reciprocal Behavior	Not Resettled ($n = 513$)	Resettled ($n = 246$)
Money Loaned		
Percentage of households that loaned money	10.5	8.9
Average number of households lent to	1.59	1.45
Amount of largest loan (yuan)[†]	11,339	3,273
Money Borrowed		
Percentage of households that borrowed money	10.8	6.5
Average number of households borrowed from	1.84	1.45
Amount of largest loan (yuan)[†]	2,709	10,121
Labor Provided		
Percentage of households that provided farm help	56.5	62.2
Average number of households helped[*]	3.04	2.35
Average person-days of labor provided per month[†]	5.46	4.58
Labor Received		
Percentage of households that received farm help[†]	58.9	68.7
Average number of households providing help[*]	3.01	2.40
Average person-days of labor received per month[†]	5.57	4.75

Note: Independent-samples t-tests and chi-square tests (* significant at 0.05 level; † significant at 0.01 level).

and chi-square tests, to determine whether there were significant differences in reciprocal behavior based on resettlement status.[12]

The most striking finding shown in table 3.4, and one that may have long-term implications for these communities, relates to borrowing and lending activity. Most borrowing and lending take place within immediate or extended families, between parents and their grown children, or among members of the kinship networks of the household head or spouse. Households from the nonresettled category were net lenders; they loaned out far more money than resettled households, and their largest reported loan (11,339 yuan on average) was three times higher than that reported by resettled households (3,273 yuan). By contrast, resettled households tended to be net borrowers. Although a comparatively small number of these households had borrowed money during the year in question, the average amount of the largest loan they incurred was 10,121 yuan, which amounts to about one-quarter of the annual income from all sources. In short, resettled households often borrowed fairly large sums of money. It is difficult to say with certainty, but this heavier borrowing could be indicative of a trend of indebtedness for households dealing with the shocks of displacement and resettlement: they borrow money to temporarily offset lost income or to invest in making improvements, in irrigation canals or other infrastructure, on their newly acquired agricultural land.

Nonresettled households typically enjoyed more robust labor-sharing networks, both providing and receiving labor with a larger number of kin, friends, or neighbors. Most labor sharing takes place within the village, and almost all of it takes place within immediate or extended families. The vast majority of such activities occurred during the busy season, between the spring and autumn equinoxes (*chunfen* and *qiufen* on the Chinese agricultural calendar). Much of this work was devoted to rice production, including turning over the soil, leveling and flooding the paddy, cultivating juvenile rice shoots in the germination nursery, transplanting shoots into the fields, weeding, monitoring irrigation, and harvesting and threshing the rice. Tasks involved in maintaining dryland crops can vary depending on the species composition of the field but generally include planting, irrigating (often by hand), weeding, applying fertilizer, and harvesting.

Do villagers' subjective attitudes about their communities differ by resettlement status? This portion of the survey was based on standardized, previously validated measurements regarding sense of community (Van Deth 2003), adapted to fit the local context in rural Yunnan. It consisted of seventeen items that assessed various dimensions of community, including feelings of trust, common identity, satisfaction, and attachment. The results are displayed in table 3.5.[13]

Somewhat surprisingly, resettled villagers reported higher levels of community satisfaction on many of the scaled survey items. Resettled villagers were more likely to trust their neighbors in an emergency situation; to report higher levels of satisfaction and happiness; to express positive feelings about how their community has changed over time; and to feel optimistic about their community's ability to solve various problems as they arise. A word of caution is in order when interpreting these results, however. Because of the cross-sectional study design, it is not possible to show that resettlement *caused* the differences seen in the data. Correlation, however, can be clearly seen, and there is reason to suspect that these communities, having recently undergone some major socioeconomic shocks due to resettlement, have found ways to adapt by bonding together and relying upon one another's support.

What are the long-term implications of these trends for households living in the Lancang basin and facing the environmental and social changes wrought by dam construction? The answer to this question requires a basic grasp of the changing terrain of agrarian political economy in postreform China. In the mid-1940s, the anthropologist Fei Xiaotong and his student Zhang Zhiyi described the household-based system of economic production that formed the backbone of Chinese society for millennia before the Communist Revolution: "The household is the social and economic unit, holding property in common, living together in the same house except for temporary absences, and working under the system of the division of labor to keep life going" (Fei and Zhang 1945:11).

When these words were published, Fei likely had little idea that the traditional system of smallholder farming would soon be turned on its head. After the establishment of the PRC in 1949, the CCP gradually established a system of governance based on socialist ideals. By 1952, private farming was abolished, and a system of collective agriculture involving communes, brigades, and production teams was instituted; this system became the basis for economic and social life in rural China for nearly three decades.

TABLE 3.5 Comparison of Villagers' Attitudes Between Resettled and Nonre-settled Communities

Statement	Percentage Agreement	
	Not Resettled (n = 513)	*Resettled* (n = 246)
1. I feel loyal to other villagers.	87.3	92.3
2. I feel that other people in this village are like me.	83.8	82.9
3. If I need an opinion about something, I would ask people in this village.	78.0	70.7
4. I trust that people in this village would help me in an emergency situation.[†]	85.0	94.7
5. I would cooperate with others to improve this village.	92.8	95.5
6. I would prefer to stay in this village rather than move.	51.5	59.3
7. On the whole, I like living here.	70.6	76.4
8. On the whole, I am happy.[†]	66.7	78.0
9. I frequently stop to greet other villagers.	86.2	82.1
10. Neighbors often come over to my house.[†]	76.7	85.4
11. I often exchange things and help with my neighbors.	93.8	91.5
12. People in this village return money they borrow.	88.3	87.0
13. On the whole, I feel I can trust most people in the village.	81.8	85.8
14. On the whole, the level of trust in the village has improved over the past five years.[†]	57.1	68.7
15. Compared with other villages, this village has fewer problems.[†]	51.9	63.8
16. If there were a government program that did not benefit my family but benefitted the community at large, I would support it.	96.1	94.3
17. People in this village can usually solve their own problems.[*]	59.5	69.1

Note: Chi-square tests (* significant at 0.05 level; † significant at 0.01 level).

After Deng Xiaoping ascended to power in 1978, the party began experimenting with decollectivized agriculture in several provinces, and when these experiments met with initial success, agricultural collectives throughout China were rapidly dismantled. By the 1980s, peasants were subject to the new Household Responsibility System, under which they were granted land leases on five-year terms, which were extended to fifteen years in 1984 and thirty years in 1993. Under the Household Responsibility System, peasants are free to make crop-selection decisions and to sell crops on the market for profit after meeting basic grain-procurement requirements set by the state (Oi 1999; Selden 1998). This effectively reinstituted China's long-standing tradition of smallholder agriculture; all decisions regarding crop selection, cultivation, and market distribution are made at the household level, and economic risk within China's rapidly changing market economy is also assumed at the household level.

Security of future land-tenure rights is a major issue in the Lancang basin and elsewhere in rural China. In contrast to urban land, which is owned by the central government (*quanmin suoyou zhi*), rural land rights are vested in rural collectives (*jiti suoyou zhi*) at the level of township, village, or production cooperative. Individual households are granted certificates that give them use rights but not full ownership rights over two types of land leased from the rural collective: "responsibility land" (*zeren tian*) and "contract land" (*chengbao tian*). Responsibility land is allocated to households in exchange for delivery of a grain quota to the state, whereas contract land is leased to households who wish to expand their land holdings, often through a bidding process (Swinnen and Rozelle 2006:57; Rozelle et al. 2005). The collective maintains the right to appropriate land within its jurisdiction when necessary. Agricultural land may not be bought or sold by individuals, and those who have invested in infrastructural improvements such as irrigation have no guarantee that they will ultimately benefit from such investments. The current smallholder system is thus the product of a combination of factors, including ecological and topographical conditions that make large-scale, mechanized agriculture difficult, central policy favoring collective land ownership, and a complex leasing system that makes consolidation practically impossible.

Viewed in the light of China's Reform and Opening policies, the shift to commodity crops is a rational response to the profit motive created by market liberalization: grain prices remain under the control of the central government, but nongrain commodity crops prove a lucrative option for farmers who can gauge market demand and adjust their cultivation practices accordingly. Most farmers welcome any increase in cash income because the central and provincial governments continue to implement Reform and Opening policies by pulling back from the provision of key services such as education and health care. The continuing liberalization of China's agro-economy induces most farmers to participate in the market in order to provide a decent standard of living for their families.

However, market-driven agriculture under the Household Responsibility System creates considerable risk for individual farming families, who must meet their own economic needs as the state provides less security and fewer services than it had during the socialist period. High-value commodity crops, such as walnuts, are not a viable strategy for displaced households. Such households tend to make up for this deficiency by relying on government compensation and by participating in a wide array of wage labor and entrepreneurial activities. They also depend on the support and reciprocity of their extended families and their neighbors.

LIFTING THE LID

In the summer of 2009, while conducting research in Fengqing County, our survey group traveled to a village just upstream from Xiaowan Dam. Construction on the dam had largely been completed by that point, and the reservoir was partially filled. Many villages were already under water, their inhabitants living in the migrant village in the township center. The rain that day was ferocious, coming in sideways sheets of water, and debris from numerous small landslides blocked the road in some places. We parked our van on the shoulder of the road and began a long hike down a winding, muddy path to the village; the footing was treacherous and slow going, and our group was passed several times by local villagers wearing old-fashioned palm-fiber capes as rain slickers.[14]

After half an hour of hiking, our guide, a local Yi man, received a call on his mobile phone, and the tone of his voice quickly turned dark. The call was from the village mayor, who was on his way down the trail to retrieve us, accompanied by another village official and a representative from Huaneng Corporation. After a short wait under the cover of trees, the men met us on the trail, informing us that we couldn't proceed to the village because it was "unsafe." Accustomed to dealing with bureaucratic obstruction disguised as concern for public safety, we persisted, asking whether we might be able to survey households higher on the mountain and away from the reservoir catchment area. The officials, however, were adamant that we had to turn around.

Disappointed and soggy, we returned to the main road, where the vice county head, wearing a police rain slicker and flanked by two Public Security Bureau officials, met us. After ensuring that our documentation was in order, he told us about the tragic events of the previous night and our unfortunate timing that had coincided with them. On July 20 at 3:00 a.m., a landslide had occurred in a hamlet adjacent to the river. At least 900 cubic meters of mountainside had sloughed off into the river, which was already swollen as the reservoir filled and the monsoon rains poured down, causing a surge that wiped out several houses. Subsequent television and newspaper reports on the tragedy said that two victims were found dead within several hours, but twelve people were still missing and presumed dead, eight from a single family (*Yunnan Wang* 2009). Within a few days, the media reports concluded that "heavy rains" were to blame for the landslides, and the thread of news soon disappeared without confirming the fate of the missing victims.

People in China often refer to any attempt to peer inside a complex and poorly understood phenomenon—in particular one involving political sensitivity—as "lifting the lid" (*jie gaizi*). In this case, the immediate difficulties placed in our path by bureaucrats represented one obstacle to lifting the lid, but it was not the only one. The effects of involuntary displacement on individuals and communities are exceedingly complicated and difficult to understand. Many questions remain only partially answered, and the long-term future prospects of displaced villagers are unclear.

Another fact that thwarts efforts to "lift the lid" on resettlement problems is that many of the long-term effects of resettlement defy easy measurement. Beyond changes in income, agricultural practices, and housing conditions, how do villagers experience the cultural and emotional dimensions of what they are going through? What are the implications, for example, of relocating a population with generational ties to the landscape, with family tombs lining the edges of their agricultural fields, and with cultural and even spiritual attachment to place? Anthony

Oliver-Smith, who has examined a range of case studies on resettlement for large-scale development projects, concludes: "Attachment to place may transcend the unique experiences of individuals and come to involve the constellation of social relations, and the cultural values that inform them, of entire groups or communities. . . . The feelings, memories, ideas, values and meanings associated with everyday life in some setting become a dimension of a person's or group's identity" (2010:166).

In my experience, many villagers felt a sense of ambivalence toward the Lancang dam projects, recognizing elements of loss while maintaining a tenacious sense of hope for the future. One middle-aged Yi farmer, whose family had recently been displaced by the Xiaowan Dam, reported that his family had moved into the resettlement village and had been fairly well compensated for their lost land. He reflected, "The biggest effect on us is the inundation of our land. The dam will flood farmers' houses and fields, and they will have to move. Many have already been moved. This destroys people's traditions and culture. People are usually not willing to move. It totally changes their way of life. Of course, that's progress [jinbu]. If the dam wasn't built, there would be no social progress [shehui jinbu]. There are more benefits than drawbacks. Of course, there are some negative impacts. I lost my land. But the positive benefits outweigh the negative impacts. Life is better than before."

His remarks underscore the fact that as the Lancang dams move forward, displacing tens of thousands of villagers, they change people's sense of long-term security; the prospect of losing any amount of farmland, which is the primary source of livelihood and social security in rural China, can be terrifying for farmers. At a basic level, the dam projects are redistributing economic risk among rural households, exposing people to considerable social and economic vulnerability. The past several decades of economic reform have unraveled many of the previously foundational social institutions in rural China, such as the family and the agricultural collective. Many of the reforms of the 1980s and 1990s, for example, were seen not only as "Reform and Opening" (Gaige Kaifang) but also as "untying" (songbang): untying the villager from the collective, untying the economy from the central plan, and untying the individual from his or her natal community. Ironically, of course, the party–state has engineered and directed the policy changes that allow for such transformations of personhood.

The "untying" of households from the collective agro-economy has left many farming households feeling economically well off as the beneficiaries of "social progress," but it has also left them vulnerable to a new range of economic risks. Many villagers had made a calculated decision about which crops to grow and whether to focus on subsistence or market sale; they were responsible for distributing their crops; and they needed to pay cash for medical treatment or educational expenses for family members when the need arose. They adapted to these changes by sending household members into the labor market, starting new entrepreneurial ventures, or borrowing money from neighbors and family members to get by.

Toward the end of our time conducting surveys in Fengqing County near the Xiaowan Dam site, our research group took a detour to spend nearly three hours winding along cobblestone roads in order to view a regionally renowned tea tree that locals claimed is 2,000 years old. At the tree—which was indeed the largest and most impressive specimen I had ever seen, notwithstanding the dubious claims about its antiquity—we met a group of middle and high school students from Xiaowan Township. One of the boys, a short kid with glasses, was headed to university that fall; he had scored exceptionally well on the entrance exam and earned a place at Tsinghua University in Beijing, the country's top institution for science and engineering. During the course of small talk, a member of our group asked him, "What do you plan to study?" Without a moment's pause he replied, "Water resources and hydroelectric engineering [*shuili shuidian*]."

4

THE NU RIVER

Anticipating Development and Displacement

FROM THE small city of Baoshan in western Yunnan, it is an hour's drive northwest to the Nujiang Lisu Autonomous Prefecture. From Liuku Township, at the southern end of the prefecture, the road traces the Nu River Gorge almost due north for 300 kilometers to the county town of Gongshan and, beyond it, the Tibetan frontier. The Burmese border lies just 20 kilometers to the west, up and over the jagged spine of the Gaoligong Mountain Range. Small tributaries flow down into the main stem of the Nu at regular intervals; over time, these streams have built up alluvial fans of rich soil spanning hundreds of meters along the riverbank, which villagers have terraced and irrigated for rice production. These tributaries, along with abundant local springs fed by groundwater, also provide the supply of drinking water for local villagers.

The Nu River, its turbulent water tinted a glacial blue gray, flows swiftly by. Rice paddies crisscrossed by narrow irrigation canals line the lower areas of the gorge near the river, giving way higher upslope to rows of corn and other dryland crops and, finally, to mixed deciduous forest. Through low cloud cover, small hamlets can be seen perched on ridgelines high above the gorge, a day's hike from any road. The Nu River Gorge is strikingly beautiful, but the people who make a living here, largely by farming and scarcely beyond the level of basic subsistence, are startlingly poor. When our research team first began to compile the data on household income for local villagers, one of my students constructed a spreadsheet that showed the basic trends across our various research locations. At first glance, I was sure that he had misplaced a decimal for the Nu River data and that the actual income figures must be an order of magnitude higher; careful checking of the data, however, confirmed that the figures were unfortunately correct.

Life is changing rapidly for people in the Nu River Gorge. For the past two decades, tourists in search of unique ecological and cultural experiences have beaten a path to northwest Yunnan. But they have stayed largely east of the gorge, on the well-worn circuit from Kunming to Dali, Lijiang, and occasionally Shangri-La. However, the upper reaches of the Nu watershed, beyond the county town of Gongshan, are currently undergoing rapid infrastructural development that will allow road passage from the Nu River Gorge eastward to Deqin, Zhongdian, and Lijiang, which are already popular tourist destinations. In the midst of this construction boom, the subsistence economy based on crop production and wild-resource harvesting is gradually giving way to wage-labor jobs, migration to cities and towns, and tourism-related work. These changes are commonplace in many picturesque locations in rural China as middle-class people with newly acquired disposable income seek novel tourist experiences; but in the Nu River Gorge, villagers also must cope with a hydropower-development scheme involving more than a dozen dams that has spurred years of controversy.

In contrast to the Lancang basin, where dam development is quite far along, the pace in the Nu basin is much slower, and development is proceeding in fits and starts. Villagers who live along the Nu River, among the most economically and culturally marginalized people in the nation, face an uncertain future as they seek to understand the hydropower-development plans for their region and to anticipate the effects on their lives.

GEOGRAPHY, ECOLOGY, AND ETHNICITY IN THE NU RIVER GORGE

Like the Lancang, its neighbor to the east, the Nu River originates at more than 5,500 meters high on the Qinghai–Tibet Plateau. But the Nu cuts a straighter path southward through Yunnan for several hundred kilometers, carving through the deep gorges of the Gaoligong Mountain Range. Downstream, the river continues its course through Myanmar, where it forms part of the border with Thailand and is known as the Salween (refer to figure 1.1 in chapter 1). The segment of the river in Yunnan encompasses a variety of ecosystems from north to south, including

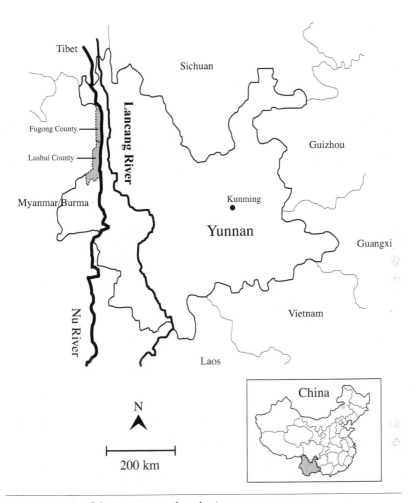

FIGURE 4.1 Map of the Nu River and study sites.

glacial scree, alpine meadow, alpine conifer, deciduous forest, pine forest, mixed forest, savannah, and riparian habitats (Xu and Wilkes 2004). For most of its course in Yunnan, the river flows almost due south, flanked by 4,000-meter mountains on either side. Ferns, lichens, and epiphytes, many of which the local people gather as food, thrive in the foggy depths of the gorge, as do many species of bamboo, ornamental orchids, and camellias.

Two early Western explorers, the Briton Francis Kingdon Ward and the Austrian-born American Joseph Rock, made excursions into the Nu River Gorge in the early years of the twentieth century, leaving detailed journal accounts that have since been published. Although they shared common goals and experiences in the course of their explorations, their rhetorical styles could not be more different. Both men focused their attention on geography and ecology, with some intermittent detours into the ethnology of local people, but Ward's depictions of the region are highly personal and reflective, containing flowery prose bordering on the poetic, whereas Rock's style is spartan, no nonsense, giving the reader a day-by-day description of his travels and little sense of introspection.

The written accounts left by these early explorers give the reader a feeling for the isolation and uniqueness of the Nu River valley. Traversing from the Mekong to the Nu, across the spine of the Khawa Karpo range and along a tributary, Ward describes his first encounter with the river in 1913:

> Over the top of the ridge, only a few hundred feet high, we see huge portals framing the entrance to a wild rocky valley, lying south-west, and it is here that the river [Wi-chu, a tributary of the Nu] finally wriggles its way out of this mountain maze and turns bravely to meet the Salween. In the failing light, the view down the valley, girded by giant rock ribs, is wildly imposing. Terrific gusts of wind buffet us in the face as we continue down the river, and I find the greatest difficulty in taking compass bearings. Now the pink glow which for an hour has lingered, changes to silver as the moon rises into a sky of palest blue, illuminating bands of white road. I feel in the highest spirits, and sing as we march along in the warm darkness under the brilliant dome of night.
>
> (1923:162)

Joseph Rock, who resided on the outskirts of Lijiang from the 1920s until national liberation at the hands of the Communists in 1949, published a huge collection of photographs and narratives in Western periodicals such as *National Geographic* in addition to numerous scholarly papers on botany and ecology. He also penned a book on the geographic and cultural history of the region, entitled *The Ancient Nakhi Kingdom of Southwest China.* In that volume, Rock reflects on the remarkable fact that a comparatively short trek allowed one to traverse three tremendously varied watersheds:

> The high mountain ranges which separate the Salween, Mekong and Yangtze—to mention the three longest rivers only—all reach their greatest height at about the same latitude (28° 20'); they act as rain screens, collecting the monsoon clouds which sweep across from the Assam plateau and assure a plentiful rainfall. To the north of the mighty snow peaks the land is parched and arid, a rocky waste. A rope could be stretched between the rain belt and the arid zone, so close do they adjoin each other. . . . The heaviest rainfall, as well as the first snow, is caught by the Salween–Irrawadi divide. . . . Next comes the Mekong–Salween divide, called the Kha-wa-kar-po Range, 21,000–22,000 feet in height and distant only 36 li from the Salween as the crow flies. While both these ranges and their passes are covered with snow by October, the Mekong–Yangtze divide, which reaches its highest point in Pai-mang Shan, about 20,000 feet in height, is still free of snow in December, and several species of Gentians enjoy a glorious sunshine and blue sky. Thus do the western ranges gather all the moisture.
>
> (1947:276).

In the writings of both men, one can sense a mixture of wonder, bewilderment, and sometimes open condescension toward the human inhabitants of the Nu River Gorge. Upon meeting the Nu and Lisu, the two predominant minority groups in the gorge, Joseph Rock observed matter-of-factly, "Nearly all are cretins and afflicted with goiter. Their constant intermarriage has caused them to degenerate. They live a miserable existence, isolated from the rest of the world" (1947:426). The word *cretin*, which offends modern ears, did not carry quite the derogatory connotation of today; rather, he used the term to refer to people's small stature, which he attributed to underactive thyroid glands, probably due to malnutrition from a diet lacking in iodine. Goiter—a swelling of the thyroid gland that can cause noticeable bulges in the neck—is a symptom visible in many of the people posing in Rock's photographs.

Joseph Rock's early explorations were commissioned by the U.S. Department of Agriculture, with the intention of discovering blight-resistant varieties of chestnut trees for introduction to North America. Many of his scientific contemporaries considered him a "walking encyclopedia"

(Harrell 2011:7), and he collected more than 100,000 plant specimens during the years he spent in East and Southeast Asia, naming and cataloging an incredible list of species from azaleas and rhododendron to peonies, poppies, and ferns (Yoshinaga et al. 2011:123). But the scope of his work turned out to be much wider than botany: he contributed to the development of what is arguably the most comprehensive dictionary of the Naxi language, a multivolume tome that was published posthumously in 1963 and 1972 and has become a cornerstone of Naxi language revitalization efforts. When Lijiang Old Town became a UNESCO World Heritage site in the 1990s, attracting millions of tourists from around the world each year, it was through Rock's writings that much of the Western world first came to know northwest Yunnan.

In Francis Kingdon Ward's work, the reader can also see a struggle to understand the intricacies of ethnic identity in northwest Yunnan. A mélange of people make appearances at various points in Ward's writing. There are the "Nung" (Nong). There are the "dwarf Nung," which are probably the Dulong because their main distinguishing feature, apart from their small stature, is prominent facial tattooing of women, a cultural practice that continued well into the mid–twentieth century. There are the White Lisu, living mostly in the lower portion of the watershed near the town of Liuku, and the Black Lisu farther north, which Ward distinguished according to various degrees of "uncouthness" (1923:176). There are the Tibetans, who lived primarily in the northern reaches of the river beyond Gongshan.[1]

Nujiang Prefecture, where eight of the thirteen dams in the Nu River cascade-development plan are to be built, is the heart of the Nu River Gorge. It was established in 1954 as an "autonomous prefecture" for the Lisu people, whose language is Tibeto-Burman in origin and has about 600,000 speakers in southwest China and thousands more in Myanmar and Thailand (*Ethnologue* 2009). Most Lisu communities are located in mountainous areas from 1,500 to 3,000 meters above sea level. As with many of the minority nationalities in the southwest, the Lisu practiced slavery into the early twentieth century (Miller 1994). Tibetans also took slaves, often from local Dulong communities in the upper Nu River watershed (Gros 2011).

But the Lisu live side by side with myriad other ethnic groups. A traditional Lisu folk tale provides a colorful origin story for the region's minority inhabitants. According to the story, in the dim mists of an indeterminate past era, a mythical brother and sister survive ninety-nine days and nights of torrential flooding by hiding away in a gourd, floating about the landscape and witnessing the destruction of the rest of human-kind in the rising floodwaters. Lacking other alternatives and persuaded by a pair of dazzling, golden-colored birds who possess the ability to talk, they reluctantly marry one another. In time, the woman gives birth to twelve children, six sons and six daughters, each of whom marries a sib-ling and goes off in search of his or her livelihood and destiny. The couple who travel north become the progenitors of the Tibetans; the couple who travel south become the Bai; the couple who go east are the Han; the cou-ple who go west, into present-day Myanmar, are the Keqin; the couple who settle on the banks of the river are the Nu; and the dutiful couple who choose to stay with their parents are the Lisu (Miller 1994:74–84).

Such a story provides a basic framework for making sense of the intri-cacies of ethnic identity in this corner of southwest China, but it also obscures the fact that ethnicity in the Nu River valley is often not so clear-cut. Individuals from different ethnic backgrounds have been connected via caravan routes up and down the gorge for centuries and have often intermarried. During fieldwork in the gorge, when I inquired about a person's ethnic identity, I often heard a long pause, followed by an intri-cate recitation of genealogy: "My father is Lisu, but my mother is Tibetan, and many members of my extended family are Nu." As a result of this prolonged, intimate contact between ethnic groups, facility in two or three languages is common among local residents. Cross-border transit into Myanmar, although substantially more difficult in recent years, is still fairly common; many young men from local villages are involved in the clandestine import of Burmese jade, which is highly prized in the Chinese market.

Further complicating the story of ethnic identity is the fact that Tibetan Lamaistic Buddhism as well as more ancient forms of animistic belief, long the dominant religious traditions of the area, were supplanted by Roman Catholicism when a group of hardy French missionaries began proselytizing the area beginning in the 1850s. Their efforts proved suc-cessful, particularly among the Lisu, who converted in large numbers; dozens of rustic churches now dot the Nu River valley, their crosses con-spicuously visible from the main highway. Mass is held in several lan-guages at least once each month, often at the hands of itinerant priests who travel up and down the gorge, and important life-cycle rituals such

FIGURE 4.2 A Lisu farming family's house on the banks of the Nu River.

as weddings and funerals are by and large Catholic affairs (Clark 2009).[2] Catholic churches have also come to serve as de facto community centers where villagers gather to share a meal, catch up on local current affairs, and even play a game of basketball on an outdoor court.

I interviewed one Lisu couple with five children who lived on the right bank of the Nu River in Fugong County. The Planned-Births Program (Jihua Shengyu) stipulates a maximum of two children for residents of the county, and the couple had paid a fine of 1,450 yuan for each additional child, borrowing money from their extended family.[3] Their house consisted of two rooms: a small kitchen with a dirt floor and a hearth where meals were cooked over a wood fire and a larger bedroom with a cement floor and several wooden beds covered in mosquito netting. A satellite dish sat outside the front door, connected to a small television in the kitchen, which the children watched from low wooden stools. The roof was constructed of corrugated metal, a substantial improvement over the leaky slate roofs on the houses of villagers of even more modest means. The father, a strong, compact man in his forties, showed me his most prized possession: the family Bible, written in the Lisu orthography that had been invented by Christian missionaries

FIGURE 4.3 Villagers cross the Nu River by zipline.

in the 1930s.[4] He expressed a measure of pride in his ability to read Bible verses to his children.

Like many of the highland ethnic groups in southwest China—including their closest relatives, the Keqin of Myanmar—the Lisu historically practiced swidden agriculture with minimal inputs, long fallow periods, and communal land tenure managed through kinship networks. They also relied on the harvest of wild plants and hunted the abundant wildlife of the gorge. But the rise in population over the past several hundred years, compounded by the market-based agricultural reforms of the past several decades and recent government policies severely restricting forest burning, induced them to practice sedentary agriculture. Most households now grow a variety of grain crops, including rice, corn, barley, and buckwheat, as well as a wide array of vegetable crops in small household garden plots. Agricultural fields are carved out of incredibly steep hillsides or situated on alluvial fans where tributaries to the Nu River have deposited rich, fertile sediment. Maintaining this cultivation pattern requires a tremendous amount of effort, which the anthropologist Marco Clark has referred to as a "delicate balance between gravity and human determination" (2009:23). Fishing is a common pursuit, often using a dip-netting technique peculiar to the Nu River area, involving two bamboo poles with a net suspended between (J. Goodman 2012:18).

Hard work is required even for the most prosaic of tasks. Outside the county towns of Lushui, Fugong, and Gongshan, few bridges span the Nu River Gorge, and most villagers cross the river by zipline cables or suspended footbridges to transport their crops or livestock to market. It is not uncommon to see sheep or cattle nervously thrashing about as they ride a zipline from one river bank to the other.[5] Reaching the remotest villages, which are located high in the mountains and are not visible from the gorge, requires more than a day's hike from any road.

DAMMING THE NU RIVER: UNCERTAINTY IN THE FACE OF DISPLACEMENT

In 2002, when the State Electric Power Corporation was disbanded and its assets distributed among various corporations, giving the Five Energy Giants responsibility for power generation, China Huadian Group was granted a monopoly on the right to develop dam projects on the Nu River.[6] Through its subsidiary, Yunnan Huadian Nujiang Hydropower Development Company, China Huadian Corporation enjoys close ties to the provincial government in Yunnan and exerts considerable influence over water-resource development in the region (Magee and McDonald 2009).

Current plans call for a thirteen-dam cascade on the Nu. Information on each of the thirteen dams is provided in table 4.1, from Songta Dam in the north, which is located in the Tibet Autonomous Region, to Guangpo Dam in the south, near the border with Myanmar. Design and operation specifications differ considerably from one dam to the next. Some dams, such as the Songta and Maji, are to be very large (more than 300 meters high) and will include reservoirs with enormous storage capacities, displacing thousands of people; others, such as the Bingzhongluo and Liuku, have been designed as run-of-the-river dams with minimal storage capacity and will thus displace far fewer people. The hydropower potential of the total cascade is estimated at 21,000 megawatts, larger than the total capacity of the Three Gorges Dam (Magee and McDonald 2009). Current estimates suggest that more than 50,000 people would be displaced should all thirteen dams be built, an unlikely proposition given the level of controversy now surrounding the projects.

TABLE 4.1 Design and Operation Specifications for the Thirteen Dams in the Nu River Projects

Dam Name	Height (m)	Installed Electrical Capacity (megawatts)	Reservoir Storage Capacity (million m³)	Estimated Population Displaced
Songta	307	4,200	6,312	3,633
Bingzhongluo	55	1,600	14	0
Maji	300	4,200	4,696	19,830
Lumadeng	165	2,000	664	6,092
Fugong	60	400	19	682
Bijiang	118	1,500	280	5,186
Yabiluo	133	1,800	344	3,982
Lushui	175	2,400	1,288	6,190
Liuku	36	180	8	411
Shitouzai	59	440	700	687
Saige	79	1,000	270	1,882
Yansangshu	84	1,000	391	2,470
Guangpo	58	600	124	34
Total		21,320	15,109	51,079

Note: Dams are listed from north to south.

Source: Magee and McDonald 2009; He, Hu, and Feng 2007:147–148.

To understand the perspective of local community members on hydropower development in the Nu River basin, our group of U.S. and Chinese researchers conducted household surveys in the Nujiang Lisu Autonomous Prefecture in 2009, including two counties—Fugong and Lushui—that encompass thirteen townships and twenty villages (see table 4.2). Our sampling frame was established to include both upstream and downstream communities related to four proposed dam sites—Maji, Lumadeng, Yabiluo, and Lushui—with a total sample size of 406 households (Brown and Tilt 2009). We asked households to provide information on a range of issues related to income, livelihood activities, ethnic

TABLE 4.2 Basic Characteristics of the Nu River Study Sample

County	Household Count (Sample %)	Dam Sites	Ethnic Composition	Literacy Rate
Fugong	197 (48.5%)	Maji Dam Lumadeng Dam	Lisu: 92% Nu: 6% Other: 3%	55%
Lushui	209 (51.5%)	Yabiluo Dam Lushui Dam	Lisu: 95% Bai: 3% Other: 2%	61%
Total Sample	406 (100%)			

Note: Other ethnic groups represented in the sample include Han, Bai, and Yi. Percentages have been rounded to the nearest whole number. "Literacy" refers to literacy in Chinese.

and cultural identity, community participation, and education. In addition, the research team conducted qualitative interviews with a random sub-sample of 48 households that participated in the surveys, asking questions about villagers' knowledge of the current hydropower-development scenarios being proposed, the perceived benefits and costs of dam construction, and the means available to villagers for coping with potential changes to their lives and livelihoods.

Villagers in the Nu River basin have long faced economic and cultural marginalization. For many centuries, the Lisu and their neighbors practiced swidden cultivation as a primary livelihood strategy, supplemented by the harvest of timber and nonforest timber products such as fungi and medicinal plants. In recent years, the region's integration into the market economy has led to the expansion of agricultural fields for growing commodity crops; land-use studies based on satellite images have shown that agricultural fields have encroached on forest lands as villagers struggle to make a cash income in order to educate their children and access other social services that were previously provided by the state before Reform and Opening (Xu and Wilkes 2004).

As table 4.3 illustrates, land holdings for Fugong and Lushui Counties are similar in terms of forest land parcels. But households in Lushui

TABLE 4.3 Agricultural Livelihoods and Income in the Nu River Study Communities

		Fugong County (n = 197)	Lushui County (n = 209)
Land Holdings (*mu*)	Paddy Land†	0.8	1.7
	Dry Land†	3.6	7.8
	Forest Land	6.2	7.6
Household Income (yuan)	Agricultural Income†	27	770
	Livestock Income†	1,606	2,700
	Wage Labor	400	544
	Self-Employment*	330	1,265
	Government Subsidy†	544	215
	Total Household Income	2,907	5,494

Note: Independent-samples t-test (* significant at 0.05 level; † significant at 0.01 level). "Government subsidy" includes only poverty-alleviation subsidies (*pinkun buzhu*) from the central and provincial governments, not resettlement compensation.

County, in the southern part of the watershed, hold twice as much paddy land and twice as much dry land as their counterparts in Fugong. In its northern reaches, the Nu River Gorge is incredibly steep and narrow; the soil here is rocky, and agriculture is quite tenuous. Many households also participate in the national sloped-land-conversion project (*tuigeng huanlin*) designed to reforest key watersheds; they set aside a portion of their farmland for reforestation in exchange for cash or in-kind payments from the government, which further constrains the amount of land available for cultivation.

As for most rural residents in China, for Nu River villagers their houses are their most important source of wealth. The median value of housing among surveyed households was 30,921 yuan (almost U.S.$5,000) for the total sample—21,043 yuan for Fugong County and 39,921 yuan for Lushui County. Nujiang Prefecture remains one of the poorest regions of China. Within the survey sample, 70 percent of households had been designated as "poverty households" (*pinkun hu*), which entitled them to

TABLE 4.4 Local Knowledge of and Attitudes About Dams in the Nu River Study Communities

Question	Fugong County (n = 195)	Lushui County (n = 204)	Total Sample (n = 399)
Do you think dam construction is beneficial for China?	Yes: 183 (93.8%)	190 (93.1%)	373 (93.5%)
Did you know about this project?[†]	Yes: 68 (34.5%)	172 (82.7%)	240 (59.3%)
Do you support this project?	Yes: 176 (90.7%)	172 (84.3%)	348 (87.4%)

Note: Chi-square test († significant at 0.01 level). Villagers were asked specifically about the dam project closest to them (Maji, Lumadeng, Yabiluo, or Lushui).

receive poverty-alleviation subsidies (*pinkun buzhu*) from the central and provincial governments.

A general sense of uncertainty and ambivalence pervades the perceptions of local villagers, whose understanding of the hydropower-development projects on the Nu is quite limited. In fact, most villagers, 59 percent of the survey sample, lacked systematic information about specific plans for construction of the Nu River dams (see table 4.4). Many were unaware of the scale of the dam projects that would affect them and the timeframe within which such projects might be built. Despite the mandate for public hearings, as stipulated in China's EIA Law (Chinese National People's Congress 2002a), most villagers had not attended such hearings. For example, in Xiaoshaba village, just upstream of the Lushui Dam site, where residents had already been resettled, a majority of residents reported a general lack of information about the projects; their primary source of information was word of mouth through other villagers.

A major factor influencing knowledge about the dam projects and participation in public hearings appears to be facility in standard Mandarin Chinese, or Putonghua (Foster-Moore 2010). Most residents in the Nu River Gorge speak either one or more minority languages or a local dialect of Chinese, making it difficult to access Chinese-language materials from government authorities or the media. Geographical isolation compounds this effect; in Lushui County, located in the lower portion of the gorge, where proficiency in Mandarin is more common, most residents had at least heard about the dam projects from their neighbors or relatives or through the media. But in Fugong County, where only about half of the villagers in the survey claimed proficiency in Mandarin, only about one-third of them had any knowledge about a project that could soon displace them, flood their land, and change their livelihoods forever.

Only forty-five households, approximately 11 percent of villagers in the survey, reported that their land had been officially measured in order to determine compensation levels in the event of displacement and resettlement, mostly land in the southern part of the watershed where the dam projects were proceeding most quickly. Of those households whose land had been measured, about half reported that a hydropower company official did the measuring; most others reported that it was done by a government official. Among study participants whose land had been measured, 42 percent reported feeling that the measurement was conducted fairly.

Despite a lack of systematic information about the projects, it is clear that the national discourse about hydropower development—as a poverty-alleviation strategy, as a national-security issue, and as a symbol of modernization—permeates the Nu River Gorge. In their survey responses, the vast majority of villagers reported that they "supported the construction of dams" in general in China and supported the Nu River projects in particular. Most villagers cited electricity generation and economic growth as the most significant potential benefits that would likely come from dam construction. One man, a farmer in Lushui County, remarked, "It [dam construction] can turn a natural resource into an energy source. Flowing water becomes economic income that strengthens the nation's economy and builds the nation's prestige." Some villagers anticipated more pragmatic benefits in the short term, such as employment in construction, and regarded the dam project as a means of gaining wage-labor income without having to go outside the community to look for work.

Nevertheless, most villagers recognized that such benefits would be accompanied by significant costs. When asked about potential negative impacts of the hydropower projects, many villagers mentioned the inundation of their agricultural and grazing land and the loss of their homes.

They recognized that the potential consequences for them and their families might be felt for generations, particularly if access to land—the primary social safety net in rural China—were compromised. For example, one forty-year-old farmer in Lushui County remarked, "Although villagers will receive compensation, they'll lose their land, and that will mean their children will be landless peasants."

Villagers were also asked how they would likely cope with the effects of the dam projects on their livelihoods. On the one hand, their responses often reflected a general sense of fatalism, that "nothing could be done [*meiyou banfa*]." As one fifty-two-year-old farmer said, "Whatever the government says is what we must do." On the other hand, their responses also pointed to a sense of trust in government officials and policy makers to compensate them fairly for their lost land and homes. One forty-six-year-old woman optimistically stated, "The common people [*laobaixing*] don't have any means of coping by themselves, but I believe the government will provide a plan for us." Her remarks are in line with current research in political science that suggests that Chinese villagers often trust central-government authorities, whom they perceive as upright and just, far more than they trust local-government cadres, whom they often view as opportunistic or corrupt (L. Li 2004).

ECONOMIC, POLITICAL, AND CULTURAL VULNERABILITY

Recent studies on the social impacts of dam construction around the world suggest that displacement and resettlement, particularly by force or coercion, result in a cascade of subsequent negative impacts on employment and income, social networks, and health and well-being (Tilt, Braun, and He 2009; Scudder 2005). Different social groups experience the impacts of development in disparate ways, in part based on their vulnerability, which Ben Wisner and his colleagues have defined as "the characteristics of a person or group that influence their capacity to anticipate, cope with, resist and recover from the impact of a hazard" (2004:11). In this section, I examine three key areas of vulnerability experienced by local communities as they relate to the Nu River dam projects: economic vulnerability, vulnerability related to governance and decision making, and vulnerability as it relates to cultural autonomy.

First, from an economic perspective, the southwestern region is comparatively poor. None of the provinces in which ethnic minorities compose more than 10 percent of the population is listed among the ranks of high-income or even middle-income provinces (Wang and Hu 1999). UNDP (2008), for example, uses an HDI that includes a measure of economic productivity, life expectancy, and education to identify development needs. In a recent HDI calculation, Yunnan ranks twenty-eighth out of thirty-one provinces and administrative regions. In Nujiang Prefecture specifically, minority nationality people account for 92.4 percent of the 520,600 residents, and all four counties in the prefecture have been designated as national-level impoverished counties (*pinkun xian*), which entitles their residents to government subsidies. This is a common trend in reform-era China: measured a variety of ways, income and wealth inequality has continued to grow throughout the reform era, widening gaps between rural and urban communities as well as between individuals within communities.

Both Fugong and Lushui Counties began implementing the Household Responsibility System in 1982, dismantling collective land plots and leasing them to individual farming families. Villagers grow rice and corn, potatoes, yams, rapeseed, tung trees,[7] sugar cane, and walnuts. In household garden plots, they also cultivate smaller quantities of berries, beans, cabbages, leeks, and other vegetables and fruits for local consumption. Villagers raise oxen as draft animals to plow fields and level rice paddies, along with pigs, chickens, and goats, all of which figure prominently in the local diet. In the upper part of the watershed, in Gongshan County and beyond, villagers routinely take their cattle to high-elevation alpine pastures to forage in summer.

Although these income sources are relatively easy to calculate, Nujiang villagers also gather wild-plant resources on forested land to supplement their subsistence, and these nonmonetary economic activities constitute an important part of local livelihoods. Many villagers collect mushrooms, herbs, and various plants, including *Polygonatum*, a genus of edible, starchy plants, as well as *Auklandia* and *Gastrodia*, two genera of plants commonly used in the traditional Chinese pharmacopeia (Wilkes 2005). They also gather firewood, primarily for household heating and cooking or for local exchange. Access to forested land for gathering plants takes place under an informal management system, with families and villages harvesting resources on land over which they have enjoyed long-standing usufruct claims but no formal title.

With the marketization of agriculture and the expansion of other industries such as tourism, many households have adopted fairly entrepreneurial and opportunistic outlooks. One family in Fugong County, for example, had a permanent house in a village high in the mountains, a day's hike along forested trails from the river valley. But several household members went out in search of wage work: one son worked as a driver for tourist groups, and another operated a small retail shop along the main highway in the gorge, selling drinks, packaged snacks, and cigarettes. He brought in extra money by charging villagers a few yuan for access to a rickety billiards table, a television set, and homemade corn-mash beer.

Officials in county-, prefecture- and provincial-level government agencies are largely supportive of dam development on the Nu, hoping that these projects will improve local transportation infrastructure, further develop communications technology, provide employment opportunities for local workers, spur new investment, and even improve educational levels among the local population (X. Li 2008). In fact, many of the official government documents on the Nu River dams describe hydropower development as a "poverty-alleviation weapon" (*fupin wuqi*) in a region that is otherwise perceived by mainstream Chinese society as culturally and economically backward. The government of Nujiang Prefecture, which is accustomed to relying on central government subsidies as a main revenue source, is strongly in favor of the projects; it estimates that of 36 billion yuan per year in hydropower revenue, one billion would stay in the prefecture, effectively increasing the local revenue stream by a factor of ten (Magee and McDonald 2009:50).

But careful inspections of the planning documents reveal an interesting fact: the vast majority of electrical power will be sent eastward to coastal cities in Guangdong Province under the Send Western Electricity East policy. Thus, existing economic disparities between regions are likely to be exacerbated by the Nu River projects (Magee 2006). Although at the time of our study some local villagers anticipated job opportunities working on construction at the various dam sites, such prospects can be elusive and short-lived, if they materialize at all.

A second area of vulnerability relates to governance issues; major questions about public participation in decision making as well as about compensation for lost assets remain unanswered. Recent studies have investigated the implementation of compensation policies thus far in the Nu River basin (see, e.g., Brown and Xu 2010). On September 1, 2006, the State Council adopted the Regulations on Land-Acquisition Compensation and Resettlement of Migrants for Construction of Large- and Medium-Scale Water-Conservancy and Hydropower Projects (Chinese State Council [1991] 2006). These regulations represent a dramatic step forward from China's historically inadequate compensation structure, stipulating, for example, that resettled households are to receive sixteen times the value of average annual income in addition to compensation for housing of the "same scale, same standard, and same function."

However, as is often the case, the trouble lies in the implementation of regulations. Just upstream from the Liuku Dam site, where preparatory work is under way, a total of 144 households from Xiaoshaba Village (literally "Small Sand Bar") were relocated to New Xiaoshaba Village in 2007. Although public hearings were held in 2006, most villagers reported feeling intimidated and effectively shut out of the decision-making process. In addition, when resettlement took place, residents were required to purchase their new houses at an exorbitant price. More than a year after resettlement, no steps had been taken to allocate new farmland to the resettlers, and residents reported a lack of long-term support programs such as job training, usually considered a key component of resettlement campaigns (Brown and Xu 2010). Much of the land that will be inundated is low-lying, fertile farmland adjacent to the river—land that supports the cultivation of staple crops such as rice and corn. As the anthropologist Heather Lazrus has observed, "As much as it is social, political, and physical, vulnerability is also a matter of representation to which questions of agency are central: Who is doing the representing, under what conditions, and for what purposes?" (2009:248).

The final area of vulnerability with important implications for the Nu River relates to the issue of cultural autonomy. China is a party to the Convention on Biological Diversity, one of the key outcome documents from the 1992 UN Conference on Environment and Development, Article 8(j) of which states that each contracting party shall "respect, preserve and maintain knowledge, innovations and practices of indigenous and local communities embodying traditional lifestyles relevant for the conservation and sustainable use of biological diversity" (UNEP 1992). This stipulation is not simply an altruistic appeal to the international community to intervene on behalf of indigenous communities lacking political power. Rather, support of such communities seems at least partly instrumental: with few exceptions, global biodiversity hot spots tend also to be places with great linguistic and cultural diversity. Although the precise mechanisms are poorly understood, evidence suggests that cultural diversity and biological diversity go hand in hand, that they coevolved through adaptive processes, and that the preservation of one is linked to the fate of the other (Gorenflo et al. 2012).[8]

Furthermore, the Declaration on the Rights of Indigenous Peoples (UN Permanent Forum on Indigenous Issues 2007), under consideration since 1985, was finally adopted by the UN General Assembly in 2007 and enjoyed the support of the Chinese delegation. None of these agreements, of course, has the status of international law or treaty, and all tend to be superseded or ignored when they come into conflict with domestic law or when they stand in the way of important development plans. More tellingly, the Chinese representative to the UN Commission on Human Rights, Mr. Long Xuequn, speaking at the fifty-third session of that body in 1997, is quoted as saying that, "in China, there are no indigenous people and therefore no indigenous issues" (Erni 2008:358).

The international community has regarded this statement as a declaration by fiat that China does not have to provide its minority nationalities with the rights commonly expected by indigenous peoples. China has long had at least one delegate serving on the UN Permanent Forum on Indigenous Issues, a key advisory body to the UN Economic and Social Council. However, the delegate was nominated by the Chinese government, in contrast to most of the delegates, who are elected or appointed by the indigenous groups from the nations they represent (UN Permanent Forum on Indigenous Issues 2007).[9]

This raises a more fundamental question of what the precise meaning of *indigenous* might be. Even the UN Declaration on the Rights of Indigenous Peoples, arguably the most important international agreement on the proper treatment of the world's 300 million indigenous people, fails to provide a definition of the concept. Various bodies, including the International Labor Organization, the UN Working Group on Indigenous Populations, and even the World Bank, have formulated their own. These definitions include common elements such as descent from a group who inhabited a region prior to colonization or conquest by other groups; a degree of political autonomy and self-determination; and a collective effort to defend a shared cultural heritage (Bodley 2008).

But nearly all of the words composing such definitions—the nouns, adjectives, and verbs alike—are tainted by ambiguity. What does it mean to claim descent from a particular group? What sort of time scale should be used when discussing the occupation of land or claims over natural resources? What does it mean to maintain political autonomy or self-determination, particularly in a region such as Yunnan, where the defining characteristic has long been resistance to administrative control? And what exactly constitutes cultural heritage when culture is constantly in flux—whether through internal processes such as cultural innovation or through the exogenous forces of acculturation, assimilation, market integration, or political repression?

Making a philosophical argument about the importance of cultural preservation and autonomy in the context of dam-related resettlement is not as simple as it sounds and can place one on the slippery slope toward essentialism. In Yunnan, such discussions seem inevitably to come around to ideas of cultural preservation, cultural autonomy, or, sometimes, the rights of indigenous communities. After all, this is part of Zomia, a region where such autonomy is a defining characteristic. Following a recent presentation I made at Minzu University in Beijing on the topic of "cultural preservation and economic development," a doctoral student in anthropology offered some biting criticism:

> We should turn your title on its head and talk about "cultural development" and "economic preservation." Culture is a living organism; as soon as it stops evolving, it dies. Furthermore, there is a sense of patriarchy and cultural imperialism involved in the idea of cultural preservation. We only care about "saving" minority cultures if they are "cute" [ke ai]. No one talks about preserving Taliban culture or even Basque culture in Europe because these are powerful groups with distasteful political agendas. Cultural preservation in China requires us to view minority groups as vulnerable and cute, like children.

We might add to this critical observation the fact that in places like Yunnan the struggle to "save minority cultures" also hinges on a view of minority people as bastions of unalloyed tradition, when the truth is much more complicated. It would be a mistake to look to the people who live there as untouched strongholds of minority culture, unsullied by national or global forces of economic development. The social, economic, and cultural changes faced by Nu River villagers today—agricultural marketization, increasing participation in wage labor, and likely displacement and resettlement to make way for dams—in fact represent the extension of a long trend involving such dynamic forces from outside the region. Over the past century and a half of documented history alone, Nu River residents have seen waves of religious proselytizing, ethnic and linguistic fluidity between the myriad minorities in the gorge, and the integration into the Chinese nation-state of people who had previously given little consideration to national politics.

Special consideration of vulnerable populations should undoubtedly be part of the calculus of building dams on the Nu River. But from a strictly pragmatic standpoint, foregrounding the issue of "culture" in the debates about the future of the Nu River is a tactic with a low probability of success. Although this argument may sound peculiar for a cultural anthropologist to make, I suggest in subsequent chapters that the more general concept of "rights," of which cultural rights constitute just one subset, should be the principal concern.

PRELUDE TO CONFLICT

One of the most conspicuous signs of development in the Nu River Gorge, in addition to ubiquitous and near-constant road construction, is the installation of small-scale "diversion hydropower projects," at least twenty-seven of which have been completed from Liuku in the south to Gongshan in the north, a distance of about 300 river kilometers. Their design is ingenious, yet relatively simple: water is diverted from the main stream of the river via a cement canal or galvanized pipe, channeled along the mountainside for several kilometers on a very slight slope, then dropped hundreds of meters through turbines at a power station on the river before rejoining the mainstream.

These diversion projects are much less capital intensive than dams, which makes them an increasingly attractive electricity-generating strategy throughout the developing world (Yuksel 2007). In Yunnan, such small-scale hydropower stations tend to be owned and managed by local or regional hydropower interests largely independent from the Five Energy Giants. Because of the small scale of most of these projects, they undergo scrutiny only by county-level officials and are not subject to full EIAs. Owing to the mountainous terrain and the monsoon climate, landslides are common on slopes that have been disturbed by such projects, and their scars can be seen all along the banks of the river.

Meanwhile, the status of the thirteen major Nu River projects changes regularly with the prevailing political winds, and policy debates over the past decade have had the back-and-forth quality of a tennis match. After considerable opposition from Chinese and international conservation organizations, Premier Wen Jiabao ordered a temporary halt to all Nu dams on February 18, 2004. This was considered the first serious test of China's new EIA Law, which was promulgated in 2002. Under Premier Wen's advisement, the NDRC, along with the State Environmental Protection Administration, conducted a review of hydropower development on the Nu. Early in 2006, officials decided to allow construction to commence on a scaled-down version of the projects, a turn of events often referred to in the media as the "great adjustment" (*da tiaozheng*), beginning with four of the thirteen dams: Maji, Yabiluo, Liuku, and Saige (Chen 2006). Two of the four, Liuku and Saige, were set to begin construction during the Eleventh Five-Year Plan period (2006–2010), according to a document released by the NDRC entitled *The Plan on the Development of Renewable Energy During the Eleventh Five-Year Plan Period* (see Brown and Xu 2010).

However, plans were stalled yet again by Premier Wen in April 2009, who declared, somewhat obliquely, that authorities should "widely heed opinions, expound on [the plan] thoroughly and make prudent decisions" (qtd. in Shi 2009). Speculation abounded that Wen, who has an academic background in geology, was concerned about potential seismic hazards in the aftermath of the Wenchuan earthquake in 2008. Much of southwest China is known for seismic activity: a 1996 quake centered near Lijiang killed hundreds and caused widespread structural damage to the tourism-dependent Old Town district. However, by early 2011, top officials had again decided to move forward with the projects; Shi Lishan, deputy director of the Energy Department, was quoted in the *China Daily* as saying, "I think it's certain that the country will develop the Nujiang [Nu River]" (*China Daily* 2011). On January 23, 2013, the State Council announced that Songta Dam, the dam situated highest in the watershed and the only one located in the Tibet Autonomous Region, will officially begin construction during the Twelfth Five-Year Plan period (2011–2015). Four more dams— Maji, Yabiluo, Liuku, and Saige—were also officially approved, and China Huadian was given permission to begin construction within short order.

It is nearly impossible at this point to speculate about the final number of dams that will be approved and completed on the Nu. International Rivers (2013b), a prominent NGO whose mission is to preserve the world's free-flowing rivers, published a database in 2013 based on an extensive literature review, media reports, site visits, and communication with various experts; it reported that as many as twenty-seven dams are under some form of consideration or planning by China Huadian (in Yunnan), Datang Corporation (in the Tibet Autonomous Region), and central and provincial government authorities.

There has been little active resistance on the part of local villagers in the Nu River basin in the face of these megaprojects that threaten their livelihoods. Such reticence is due largely to the lack of information about how projects are proceeding, weak capacity to mount a campaign in the face of economic and cultural marginalization, and the high political risks involved in any opposition strategy.[10] However, as we will see in later chapters, the Nu River has become a focal point for domestic and international NGOs, which see its damming as an unacceptable cost of economic progress. In response to the State Council's announcement in 2013 that placed the Nu River projects at center stage and appeared to signal a clearing away of the last political obstacles in the path of hydropower development on the Nu, International Rivers circulated a press release that called on Chinese officials to uphold the moratorium on the Nu projects, urging UNESCO to "remind China of its obligation to protect the Three Parallel Rivers Area under the World Heritage Convention" (2013a). Several domestic NGOs—including Green Earth Volunteers and Green Watershed—have initiated public-education campaigns that highlight the environmental and social costs of the projects. After more than a decade of planning and controversy, the Nu River dams remain a focal point of environmental, social, and political conflict, even as construction moves inexorably forward.

5 |

EXPERTS, ASSESSMENTS, AND MODELS

The Science of Decision Making

ALTHOUGH MOST experts readily acknowledge the fact that dams alter ecosystems and communities—sometimes irreversibly—the science, policy, and politics surrounding hydropower development are often contentious. Such controversy arises in large part, I have learned, from the epistemological differences between experts from different fields—engineers, economists, hydrologists, geographers, anthropologists, and others who, by virtue of their training, have come to see the world quite differently. Yet collaboration, cross-disciplinary thinking, and holistic analysis are all necessary preconditions to understanding and addressing the complex problems related to hydropower development. How do scientists and policy makers reach decisions about the management of water resources for ecological benefits, for human use, and for hydropower production? What criteria enter into the decision-making process, and how are these various criteria weighted by those in a position to steer policy? In the political economy of knowledge about water resources, whose expertise carries the day?

The anthropologist Laura Nader, who has long advocated the practice of what she calls "studying up," suggests that the first step in understanding decision making in a complex institutional environment is to take a close look at scientists and other experts within the institutions where they work. Nader writes, "The point of this work is not to 'put scientists in their place' (although one might want to). . . . The point is to open up people's minds to other ways of looking and questioning to change attitudes about knowledge, to reframe the organization of science" (1996:23)

In other words, scientists and policy makers are social actors, too, and thus worthy subjects of study in their own right. The ways in which they approach their fields of study, how they produce knowledge, and how this knowledge affects policy and regulatory decisions—all are crucial, if seldom examined, topics. My aim in this chapter is to examine the epistemological processes involved in high-level decision making about water resources and dams in contemporary China. I do this by drawing upon observations and interviews with scientists and policy makers from a variety of disciplinary backgrounds and by reflecting on my own experience as a scientist engaged in a large, interdisciplinary project designed to create a computer model for understanding the effects of dams on ecosystems, communities, and geopolitical relations. Such an approach gives us a window into how different scientific disciplines, each with its own epistemology, understand dams. It involves a close look at how they value certain impacts over others, how they measure those impacts, and how they conceive of trade-offs between multiple hydropower-development scenarios.

NEW FRAMEWORKS FOR DECISION MAKING

The international politics and policies of building large dams have changed considerably in recent years, from a model in which international financial institutions such as the World Bank took the lead in financing and expertise to a much more decentralized model driven by national governments and private entities. Throughout this transformation, the hydropower industry has found itself under intense scrutiny by citizens and activists from around the world, who increasingly voice concern about the social and ecological costs of dams. The WCD, an advisory body under the auspices of the World Bank and the World Conservation Union, published a landmark study in 2000 that noted that although dams had contributed significantly to human development over the years, their deleterious impacts on social and environmental systems had long eluded meaningful scrutiny. Drawing on the expertise of many scientific disciplines and a review of case studies from around the world, the WCD released what it called Seven Strategic Priorities, areas in which both science and decision making urgently needed to be improved: (1) gaining public acceptance for hydropower projects; (2) conducting comprehensive options assessment for different hydropower-development scenarios; (3) addressing the social and ecological problems of existing dams; (4) sustaining rivers and livelihoods; (5) recognizing entitlements and sharing benefits among stakeholders; (6) ensuring compliance with the best practices in the industry; and, somewhat obliquely, (7) sharing rivers for peace, development, and security.

The WCD's work calls into question many of the standard ways of measuring and evaluating the ecological and social costs of dams that have been around for decades, including the tool par excellence, cost–benefit analysis, which has historically overlooked or severely underestimated the long-term impacts of dams on local communities. But the WCD was a temporary advisory body, not a regulatory agency; it was not endowed with any sort of enforcement authority, particularly in cases such as the Lancang and Nu River basins, where dam projects are funded by domestic government agencies or corporations. The WCD's success, limited though it has been, lies in the way that it has changed the international conversation about dams, shedding light on the ways in which scientific experts and policy makers evaluate the costs and benefits of major projects. At a more fundamental level, it has reminded us that the most important questions surrounding hydropower development involve not only technical and economic feasibility but also basic human values such as equity, justice, and rights.

In an interview about how to assign relative weights to different criteria in complex decision-making processes, Dr. Wu, an environmental policy analyst from the MEP, related a comical story about working with rural communities on environmental restoration projects. He had graduated from college in the late 1980s with a degree in ecology and a starry-eyed ambition to help solve the nation's environmental problems. Working in an advisory capacity with county and township officials in a relatively poor province in central China, he lectured them on the "value" (jiazhi) of ecosystem functions, explaining that planting trees on sloped land, for example, could prevent erosion and enhance the soil quality of agricultural land downslope. The local officials, likely annoyed by this precocious college graduate from the city, always nodded in assent but went right on doing whatever they had always done. The budding MEP policy analyst eventually discovered that terminology was getting in the way of effective communication. When he talked about the value of ecosystem functions, government officials tended to turn this vague term into two concrete questions: Exactly how much money can we get for these ecosystem functions? And how do we convert them to cash?[1]

My interview with Dr. Wu soon turned into a lengthy discussion of how different constituencies and even scientists from different backgrounds tend to approach problems from their own narrow perspectives. This phenomenon has been termed "epistemological pluralism" (Miller et al. 2008): the notion that people with different backgrounds and skill sets approach a problem differently because they see the world through different lenses by virtue of their specialized training. If decision making about dams is to become more effective, more just, and more sustainable, it must first become more transparent and more collaborative. Experts must find a way to understand and value one another's perspectives, an endeavor that is more difficult than it sounds.

ENVIRONMENTAL IMPACT ASSESSMENT IN CHINA

One statutory tool with great promise for improving environmental governance and bridging the divide between scientific disciplines is EIA. Major infrastructure-development projects in China have been subject to environmental review since the establishment of the country's environmental bureaucracy in the late 1970s, but the institutional framework for implementing EIA has been slow to take hold. The role model for China's EIA process was the U.S. National Environmental Policy Act (NEPA) of 1969, which requires any major actions that may significantly affect environmental quality to undergo either a full environmental impact statement or an abbreviated waiver process called a "Finding of No Significant Impact." In either case, proper steps must be taken to identify and mitigate the impacts on plant and animal species, ecosystem services, and human health. The act is precautionary in nature insofar as it encourages government agencies to think about preventing ecological damage rather than mitigating it after the fact.

Information disclosure and public participation are key components of NEPA in at least three stages of a project: at the time of filing a Notice of Intent, which initiates the project; during the scoping process to identity the range of potential environmental impacts and affected stakeholders; and during the comment period on the draft environmental impact statement.[2] In the decades since the passage of NEPA, more than eighty countries around the world have implemented EIA laws, which, although distinct in their particular details, are guided by common principles.

The EIA process in China took a major step forward with the passage of the EIA Law by the National People's Congress in 2002 and its implementation in 2003 (see Chinese National People's Congress 2002a). In line with the goal of sustainable development, the law requires that major construction projects undergo an assessment of their environmental impacts prior to construction. The institutional level of review—whether by the MEP or by provincial Environmental Protection Bureaus—depends on the scale of the project in question. Furthermore, the specific steps in the review process may vary according to the scale of the potential impacts of the project: if a project's potential impact is deemed to be "major," a full environmental impact report must be filed; if the potential impact is "light," an environmental impact report form is required; if the potential impact is "very small," only an environmental impact registration form is required (Stender, Dong, and Zhou 2002).

Although China has had a rudimentary framework for evaluating the environmental impacts of major projects for at least thirty years, the EIA Law now mandates that government entities base their decisions to approve or reject projects in part on the EIA report and to justify their decisions in writing (Tang and Zhan 2008). The law stipulates that EIA decisions be fair (gongping) and scientific (kexue) and that the approval process be well documented. The EIA Law contains thirty-eight articles, a selection of which appears in table 5.1.

The EIA Law has the potential to affect how large development projects go forward. In fact, the thirteen proposed hydropower dam projects on the Nu River provided the first serious test of the central government's intention to honor the status of the law: when Premier Wen Jiabao halted the projects in 2004, he invoked the EIA Law and criticized China Huadian Group and its subsidiaries for failing to conduct comprehensive reviews. To my knowledge, however, there has not yet been a systematic evaluation of how well or even to what extent the law is being implemented nationwide. Nor is there a good understanding among scholars and policy makers of the extent to which the EIA Law has actually influenced the regulatory approval of large-scale development projects.

The law's weakness stems from several shortcomings. It lacks specificity about who should conduct EIAs and exactly which government agencies should exercise which oversight capacities. This makes the EIA

TABLE 5.1 Selected Articles from the People's Republic of China Environmental Impact Assessment Law

Article	Description
Article 5	The government encourages relevant entities, experts, and the general public to participate in appropriate ways in the EIA process.
Article 7	The EIA Law applies to government entities at or above the municipal level, which includes provinces, autonomous regions, and centrally administered municipalities.
Article 10	Outlines the basic contents of an EIA report, which should include (*a*) an analysis of the environmental impacts that may occur if a given development project is implemented; (*b*) measures for mitigating the negative environmental impacts; and (*c*) a summary of the environmental impact of the project.
Article 13	Describes the formation of expert panels to review and approve the EIA report. Experts are selected from a list approved by the State Council or its administrative departments.
Article 15	After a given project with "significant environmental impacts" has been carried out, an appraisal of those impacts should be conducted and mitigation measures proposed.
Article 19	Only institutions approved by the State Council may conduct EIAs. This approval process requires certification through an examination. There should be no relationship (conflict of interest) between the institution conducting the EIA and the government entity responsible for approval of the EIA and project.
Article 24	If the nature, scale, or location of an approved project changes, a new EIA report must be submitted.
Article 29	If any government entity neglects its duty in enforcing EIA law, it will be subject to administrative punishment.

review process subject to influence from a range of parties, including local-government agencies that may be dependent upon revenue generated from a given project, or the hydropower companies themselves, which possess the financial means to sway the outcome of the review process in their favor. Lack of specificity also makes proper public participation difficult because it is often unclear which agency should take charge of outreach activities, including public hearings (Tang 2007).

At a more fundamental level, any legal statute is only as solid as the judiciary system that enforces and arbitrates it. In the United States, the birthplace of EIA, the statute is essentially a procedural stipulation: it requires agencies to show compliance with myriad steps—from the scoping of possible impacts to data collection to mandated public participation—and provides a legal foothold for dissatisfied parties to slow or stop a project in the courts by claiming that one or more steps was done improperly. In many Western countries, the enforcement lever that makes the whole system work is litigation, or threat of litigation, by various constituent groups. Such a threat can have a "trickle-up" effect: it forces agencies, scientists, and policy makers to try to predict the kinds of public concerns that will be raised and to create management alternatives to address them up front (Kaiser 2006). It also means that the EIA process, laden with political conflict and controversy, can drag on for many years or be held hostage by one or more powerful interest groups.[3]

In contemporary China, however, the judiciary system plays a much smaller role in environmental management. Although environmental lawsuits are increasingly more frequent, they are still a challenge for plaintiffs because the courts remain relatively difficult for most citizens to access, and entities such as NGOs often lack standing to file suit (Stern 2011). As part of my research on water resources and decision making, I have been working to understand the science and policy of EIA as a field of practice. How do experts involved in EIA conduct their work and evaluate its efficacy? During an interview, one senior environmental scientist named Dr. Liu, a woman who runs a respected research center at a major university, reflected: "Academics believe that EIA is effective. They apply it and try to develop new methods; they try to make useful instruments both at a high policy level and at a local level. They promote it as an effective tool of environmental management. . . . Especially after the 2003 EIA Law, environmental impact assessment can be a useful tool. It can be powerful because if a project doesn't pass the EIA, it will not go forward."

Many experts share Dr. Liu's sense of optimism about what the law can accomplish. But the day-to-day reality is mired in a complex set of bureaucratic problems. Unlike in the West, where environmental reviews are often conducted by a cottage industry of expert contractors from the public and private sector, in China the government places severe restrictions on who can perform EIAs. In 2005, the MEP released a set of guidelines called *Accreditation Methods for Organizations Conducting Environmental Impact Assessments* (Chinese MEP 2005). This ensures that only select, government-sponsored organizations, usually affiliated with prominent universities or research centers, can conduct environmental reviews and submit EIA reports. In order to do so, a given organization must possess a qualification certificate (*zige zhengshu*), which is issued only after a certain number of personnel have passed an examination administered by the MEP. This is in line with Article 19 of the EIA Law, which requires certification, and may help to ensure the quality of EIA science, but it also raises the question of independence and the potential for conflicts of interest. Dr. Liu felt that the accreditation requirements fundamentally undermine the integrity of the EIA process:

> Most of the institutes that do EIAs have extremely close ties to the Environmental Protection Bureaus [which have the authority to approve projects]. It's not based on objective criteria, but on what the government thinks of them. The more you look into current EIA processes, the more controversy you'll find. Sometimes the EIA is done just for show. The numbers may be right, but they're fundamentally misleading. At our institution, we discourage our professors from conducting EIAs because it can ruin a scientist's reputation. They're not done very well. In some cases, whole sections are copied from one EIA and pasted into another. That kind of poor quality damages reputations.

Dr. Wu, the MEP policy analyst, expressed similar concerns. Reflecting on the poor quality of many EIA documents, he told me that he had recently reviewed an EIA report for a hydropower project in Yunnan but saw multiple references to Sichuan Province in the text. He asked a few probing questions of the authors, who sheepishly admitted that they had simply used the cut-and-paste feature in their word-processing software to transfer whole sections of text from one report to another. Time constraints, budgetary limitations, and sometimes outright malfeasance on the part of researchers can undermine the entire process. Moreover, lack of institutional oversight is rampant at the highest levels of government; in extreme cases, multi-billion-dollar infrastructure projects can go from design to construction before any EIA is done. One recent high-profile example is the Longkaikou Dam on the Jinsha River in northwest Yunnan, which was initiated by Huaneng Corporation in 2007 but suspended by the MEP in 2009 after critics pointed out that Huaneng had not filed an EIA report (*China Daily* 2009). The NDRC later allowed the project to continue after its environmental impacts were revised downward (Hennig et al. 2013), although it is unclear whether the project design itself was revised.

The MEP requires EIA practitioners to emphasize what the agency calls "three synchronizations" (*san tongshi*) during an EIA process: project managers must show that they have considered the environmental implications of the design, construction, and operation of any given project. Given the sporadic enforcement record of the EIA Law, however, many experts joked with me during interviews that "three synchronizations" actually stands for "eating, drinking, and singing" in restaurants and karaoke bars. Such critiques underscore the fact that the weak points in the process are not technical in nature but stem from a lack of legal and political accountability. Several experts suggested that the system was rife with corruption (*fubai*). Dr. Wu, the MEP policy analyst, noted with a sense of irony, "The United States and other countries criticize China over human rights, and they point to minor examples like the case of [the artist and dissident] Ai Weiwei. It's not about individuals like that; it's a systemic problem. There is no accountability mechanism [*wenze zhi*]."

Despite these institutional failings, many environmental professionals are working to improve the EIA process from the inside. In my interview with Dr. Liu, the research center director, she offered several ideas for improving the EIA process and its associated outcomes: "I think we should learn from Hong Kong. In Hong Kong, the process is public. EIAs are done by experts who volunteer to do them as a kind of service, and the report is made publicly available. Here, we don't really believe the government can make the right decision. It doesn't have the expertise or the capacity. The process needs to be supervised by a third party. The existing system won't work."[4]

In my interviews with scientists and policy makers, I discovered that most were aware of the need to balance public participation with government oversight, a process that can be fraught with difficulty. Dr. Liu continued, "That kind of input from public opinion is good. But the agency ultimately has to make the decision. Otherwise, you get the NIMBY [not in my backyard] problem." Using English to refer to this common scenario in which powerful constituent groups exert undue influence over the process in order to push costs onto others, Dr. Liu conceded that the most difficult aspect of public participation was ensuring that it truly represented the public interest.

A case in point is the MWR's move toward comprehensive river-basin planning (*quan liuyu gong jihua*) in recent years, a process involving multiple stakeholders in envisioning the future of China's major rivers. This participatory process is undoubtedly a step in the right direction, but without checks and balances it can be dominated by powerful parties such as hydropower-development corporations, who often use their close ties with local and regional governments to co-opt the process. Media outlets have reported that each time one of the Five Energy Giants participates in river-basin planning, the result is a new comprehensive plan that pushes upward the approved installed capacity for hydropower on the river in question. Current planning documents on the Jinsha River, for example, call for a dam roughly every 100 kilometers, which would turn the river into one slackwater section after another (*People's Daily* 2012). A rhetorical call for public participation, in short, does not always result in optimal environmental or social outcomes, particularly in the absence of transparency and accountability. This points to a systemic problem with EIA that is common throughout the world: EIA is both an "applied science," using empirical methods and models to understand the potential impacts of a given project, and a "civic science," with a mandate to provide information to decision makers, engage the public in the process, and ultimately improve environmental governance (Cashmore 2003).

MODELS AND INTERDISCIPLINARY COLLABORATION

The EIA process is one tool for improving the environmental and social outcomes of dams, but the hydropower industry also needs holistic and transparent mechanisms that ensure stakeholders' ability to participate in decision making, which requires broad cooperation across scientific disciplines. As an anthropologist, I am interested primarily in dams' human dimensions: how they uproot communities, alter economies, disrupt social relationships, and drive cultural change. But I have learned over the years that it is impossible to separate these human concerns from the effects of dams on riparian ecosystems, on water quality and availability, and on interregional or even international relations. Sustainable solutions will be found at the nexus of many fields of disciplinary practice.

Along with a group of American and Chinese colleagues, I spent more than five years working on a decision-support tool called the Integrative Dam Assessment Model (IDAM), with funding from the Human and Social Dynamics Research Program at the U.S. National Science Foundation.[5] This work represents one attempt among many ongoing efforts—by governmental agencies, NGOs, scientists, and hydropower industry groups—to improve the transparency of decision making regarding dams.[6] When we first began discussing the possibility of collaboration, our research group realized that, despite our common interests, natural scientists and social scientists tend to look at our respective objects of study through our own microscopes; we use our narrow disciplinary training to study the effects of dam construction with fine-grained resolution, while sometimes missing the bigger picture. An ecologist, for example, may focus primarily on biodiversity or threatened species, but an anthropologist's main concern may be for displaced or otherwise vulnerable communities. Yet dams lie at the intersection of ecological and social systems, which means that an effect in one area is likely to have repercussions in others. For example, the adverse effects of dams on ecosystems, hydrology, and water quality (Poff and Hart 2002) often disrupt cultural conditions and economic institutions (Cernea 2003) and change the geopolitical relationships between communities, regions, or nations (Wolf, Yoffe, and Giordano 2003).

As a result, the impacts of dams—whether positive or negative—are not readily captured through the analytical lens of any single discipline. The research group to which I belonged—which included geographers, engineers, economists, hydrologists, and anthropologists—worked toward a comprehensive, systems-based approach built on both historic and contemporary data. One early inspiration for the research group was the WCD's Seven Strategic Priorities, particularly the call for "comprehensive options assessment." We focused our energies on supporting more informed and transparent decision-making processes related to dam development by creating a computer model that could help decision makers understand and visualize how a given dam project would affect human communities and ecosystems.

A model is a simplified representation of reality, a rubric for rendering complex information in such a way that it can be more easily understood and acted upon. Model building is therefore necessarily a reductionist undertaking. Even a map, itself a kind a model, must be scaled down to fit into its user's pocket; a map with a 1:1 scale might be highly accurate, but it wouldn't be terribly practical. The statistician George Box, reflecting on the difficulty of reducing a complex social reality to a practical model, once said, "All models are wrong; some models are useful." It is in this spirit of critical appraisal that I wish to discuss some of the key obstacles our research team faced during the modeling effort, how we sought to overcome them, and what some of the implications for decision making might be.[7] In the process, I would like to elucidate how "epistemological pluralism" played out in this particular case and how disparate knowledge systems can be brought to bear on a complex topic.

THE ANATOMY OF A MODEL

The IDAM is based on the work of the World Commission on Environment and Development, which recommended that sustainable-development projects balance the needs of biophysical, socioeconomic, and geopolitical systems. Using this idea as a foundation, our research team designed the IDAM to evaluate the outcomes of dam construction on three key systems:

- *Biophysical Impacts*: How does a given dam project affect the environmental aspects of a river ecosystem?
- *Socioeconomic Impacts*: How does a given dam project affect human communities, economies, and cultures?
- *Geopolitical Impacts*: How does a given dam project affect natural-resource decision making, governance, and interstate relations?

The model works by analyzing twenty-one different impacts, seven from each of the three systems (see tables 5.2, 5.3, and 5.4, respectively). Each impact is scored on two dimensions: *magnitude*, which is the objective measure of the severity of that impact on a 0–3 scale; and *salience*, which is the subjective importance or significance assigned to the impact, also on a 0–3 scale, by whatever group of stakeholders is using the model. Computer scientists on the team constructed a graphical user interface that allows users to input data for a given scenario and to compare different projects based on size, location, or other characteristics.

TABLE 5.2 Biophysical Impacts Measured by the Integrative Dam Assessment Model

Impact	Description
BP1: Water Quality	Reservoir may change cycling of nutrients and carbon, decrease turbidity and dissolved oxygen, or change temperature.
BP2: Biodiversity	Habitat classification of affected areas, species occurrence, changes to hydraulic habitat.
BP3: Impact Area	Measurement of surface area of the reservoir and length of river impounded.
BP4: Sediment	Reservoir may disrupt natural-sediment movement; downstream channel may degrade; depositional features and channel morphology may change.
BP5: Flow Regime	Dam may change historic hydrograph, including magnitude, duration, timing, and frequency of high and low flows.
BP6: Climate Change and Air Quality	Amount of greenhouse gas emitted, compared to emissions from coal-fueled power generation.
BP7: Landscape Stability	Weight of reservoir, distance to faults, landslide hazard, grade of slopes, erosive potential of soils.

Source: Adapted from Tullos et al. 2010.

This modeling effort represents several steps forward in the effort to improve complex, multicriteria decision making about dams. First, in line with the WCD's call for more comprehensive options assessment, it allows users to assess the effects of dams more holistically on biophysical, socioeconomic, and geopolitical systems and to see how the benefits and costs of a given dam project are distributed across these three systems. Second, it combines objective, scientific analysis on the *magnitude* of impacts with the subjective *salience* assigned by stakeholders to those impacts, making the decision-making process more equitable and transparent and allowing users to understand how and why different stakeholders might value different outcomes. In various simulation activities, we have discovered that different groups of users—government officials,

TABLE 5.3 Socioeconomic Impacts Measured by the Integrative Dam Assessment Model

Impact	Description
SE1: Social Networks	People from one community may be resettled into multiple new communities, disrupting social networks.
SE2: Cultural Change	Dams may inundate sites of cultural importance, leading to loss of traditions (including traditional ecological knowledge).
SE3: Local Electricity Access	Dams may affect connection to the power grid as well as price and availability of electricity.
SE4: Health Impacts	Reservoirs may affect health via water quality, water-borne illnesses, and availability of potable water.
SE5: Income	Inundation of agricultural land may affect incomes for farmers and change income-generating activities.
SE6: Wealth	Resettlement may affect amount and sources of wealth, especially housing and land values.
SE7: Macro-impacts	Measures the cost of resettlement, cost of infrastructure, and commercial value of hydropower.

Source: Adapted from Tullos et al. 2010.

hydropower corporation executives, and NGO representatives, to name a few—can view the exact same set of magnitude scores and yet assign wholly different salience values to the data that they see. Finally, the IDAM allows stakeholders to make explicit comparisons between different hydrodevelopment scenarios. Users can compare different watersheds, different dam designs, and even different operational characteristics to see the various impacts associated with each scenario. Once stakeholders understand the full range of impacts associated with a given dam-development project, they can begin to think more systematically about how to select the best option or how to mitigate the most serious negative impacts of a project.

TABLE 5.4 Geopolitical Impacts Measured by the Integrative Dam Assessment Model

Impacts	Description
GP1: Basin Population Affected	Proportion of basin population affected by dam.
GP2: Political Complexity	Number and type of political boundaries in a river basin may affect dialogue, cooperation, or conflict.
GP3: Legal Framework	Strong laws may mitigate the impacts of change; existing basin agreements and associated river-basin organizations may help reduce vulnerability throughout the basin.
GP4: Governmental Transparency	Openness and transparency of decision-making processes affect management capacity.
GP5: Domestic Political Stability	Cooperation during planning, construction, operation, and management phases leads to the establishment or strengthening of institutional arrangements and affects relations among relevant administrative areas.
GP6: International Political Stability	Cooperation during planning, construction, and operation, and management phases leads to the establishment or strengthening of institutional arrangements and affects relations among relevant administrative areas at the international scale.
GP7: Impacts on Downstream Parties	Dam construction results in costs or benefits for individuals and communities outside the immediate area of the dam (other counties, municipalities, provinces, countries).

Source: Adapted from Tullos et al. 2010.

COLLABORATIVE CHALLENGES

Each of the IDAM project investigators brought to the table his or her own disciplinary background, complete with an inherited set of theories, research methods, and sometimes unquestioned assumptions. The model was interdisciplinary from the beginning, requiring a fair amount of flexibility and adaptability from the research team members, a process of compromise that often proved humbling. During one meeting early in the research process, project personnel discussed a schematic chart that showed how dams are related to ecological and social changes. The causal flow of the diagram moved from top to bottom, with arrows connecting boxes. The biophysical group wanted to put the dam itself at the top of the diagram, arguing that it was a primary driver of change because it altered habitats, changed water quality, and displaced human populations. The geopolitical group wanted to put policy decisions at the top of the diagram, arguing that energy distribution and comprehensive river-basin planning were the major drivers. Meanwhile, the socioeconomic group, myself included, argued for putting human factors such as rising energy demand and anthropocentric values at the top, arguing that these factors drove policy choices, which in turn drove the biophysical reality. In the end, no single perspective is "correct"; well-intentioned experts, approaching a common problem from different disciplinary perspectives, can have widely divergent views.

One of our Chinese collaborators quoted an eight-character poetic couplet from classical Chinese that aptly summed up each researcher's propensity to view his or her own piece of the project as most important. In the couplet, a fruit vendor, competing with a dozen others in a busy marketplace, boasts of the superiority of her product: "With every melon Mrs. Wang sells, 'this one's the best,' her customers she tells" (*Wang po mai gua, zi mai zi kua*). As we sought to resolve these differences, we faced a number of technical and epistemological challenges.

Model Type

The first, and perhaps most obvious, obstacle we faced was selecting the type of model that would best represent the complex interactions between the biophysical, socioeconomic, and geopolitical aspects of dams. Several graduate students worked during the early phases of the project on a literature survey that would help us understand the range of options available to us, including agent-based models, contingent-valuation models, and others. We ran a series of pilot surveys in our study areas in Yunnan that included a contingent-valuation question—sometimes called "willingness to pay"—aimed at understanding the trade-offs between hydropower and coal-fired power plants, but the question proved too complicated and unwieldy to include in the final version of the survey. Most villagers simply did not conceptualize the trade-offs between dams and coal as an either–or proposition. Likewise, agent-based models—which are designed to predict the actions and interactions of multiple, autonomous agents in a complex system— proved inappropriate because decisions about dams are driven more by powerful institutional players such as government agencies and corporations than by individuals acting on various incentives. We ultimately decided that the IDAM would be a computational model designed to support policy decisions and that it would be interactive, allowing users to input data for magnitude and salience.[8]

Selecting Impacts for Measurement

If a model is to be successful, it must capture a complex reality and convey it in relatively simple terms. With this in mind, our team conducted an exhaustive survey of the literature on dam impacts and held a series of expert panels in both China and the United States to get input on precisely which impacts the model should measure. This process was fraught with difficulty and sometimes outright conflict; in fact, the impacts that appear in tables 5.2–5.4 are the product of years of discussion, debate, and compromise. While convening our expert panels, we routinely ran into the "Mrs. Wang phenomenon," with each expert suggesting that his or her own field was sufficiently important to warrant more comprehensive analysis within the model. To finalize the list of impacts we would focus on, we had to make hard choices, sometimes sacrificing nuance for clarity and simplicity.

For example, we called the first socioeconomic indicator (SE1) "social networks." It is designed to measure how dam-induced displacement alters the quantity and quality of social ties, which, as I have suggested in earlier chapters, can provide villagers with a means of adapting to displacement by relying on one another for financial resources or labor sharing. In order to measure it, we collected survey data on villagers' borrowing and lending networks, labor-sharing networks, participation in community activities, and subjective attitudes about the members of their community. Such choices inevitably involved a reduction of complex social realities down to a manageable set of indicators; in the process, things got left out. As an anthropologist, I would have liked to include additional measures such as psychological stress, loss of traditional ecological knowledge, and even the loss of cultural and spiritual ties to a landscape, acquired and nurtured over many generations, when communities are forced to resettle. However, facing an already large data-collection burden, with hundreds of data points for more than 1,000 households, we were forced to reduce the scope of this indicator—and others—to something more manageable.

Data Operationalization and Binning

Once a set of impacts has been agreed upon, how should we go about collecting the data to measure them? Some impacts are quite readily captured in a model, whereas others are much more difficult to examine systematically. For example, the hydrologists on our team, tasked with measuring "flow regime" (BP5), a biophysical system factor, could simply go out and collect water samples from upstream and downstream of a given dam project or even use a predictive sediment model that uses flow rate and other data to estimate sediment load. But the data-collection process can be much more difficult with complex and nuanced impacts such as "cultural change" (SE2). Once the data necessary for measuring each impact was collected, we were tasked with converting it to a scale, a process we came to call "binning." We settled on a four-point scale: 0 = no impact, 1 = small impact, 2 = medium impact, 3 = large impact. But hard choices had to be made at every step of the process. For example, if one determines that a given impact is "large," the obvious question is, "Large relative to what?" When our research group convened a symposium on the Columbia River and was touring the Bonneville Dam, the docent from the U.S. Army Corps of Engineers told us about the river's volume, the technology behind the dam's turbines, and other key facts. About one year later, we convened a similar workshop in Maine, where we toured a small flow-control dam on the Kennebec River. We all had a good laugh when one of the hydrologists in the group informed us that the flow volume of the Kennebec was about a thousand times smaller than that of the Columbia. Such field experiences underscored the importance of thinking about scale in the modeling effort and encouraged us to allow enough flexibility in the model to capture different hydropower-development scenarios, from the mammoth Three Gorges Project to more modest projects. As a result, the scientists and policy makers who use the model will need to agree upon the scale of their modeling effort as a precondition to their work.

Measuring Change

One of the problems of understanding the effects of dams is that such effects often unfold over an extremely long time horizon. The life cycle of a dam—from design to construction, operation, and finally decommissioning—typically lasts for many decades. Our research team tried for a very long time to analyze and measure diachronic "change" within the various biophysical, socioeconomic, and geopolitical systems affected by dams, but we came to realize that doing so is a tall order. For example, if we set out to measure changes in social networks in response to dam-induced displacement, we would need to collect baseline data from before construction on the dam was initiated and then collect similar data after completion of the dam and displacement. The costs of such an undertaking—in time, energy, and financial resources—would be enormous. We opted instead to use a cross-sectional study design that allows us to compare multiple dam sites at a single snapshot in time, an approach that, as I described in the Lancang case study (chapter 3), does not lend itself to measuring diachronic change.

We also encountered the related problem of cumulative impacts. When several large dams are built on the same river or even on tributaries of the same river system, the impacts—on water quality, threatened species, and human communities—accumulate. But EIAs and other scientific studies are generally conducted for single projects, not for multiple projects in the same basin. To complicate matters, no one really knows about the shape of the impact curve: Do the effects suddenly ramp up with the addition of one more dam, crossing a kind of threshold that causes irreversible perturbations in the system, or do the effects accumulate more gradually? Does the first dam built in a basin cause significant impacts, while the effects of subsequent dams are only marginal?[9] As it turns out, these crucial questions and many more just like them remain unanswered.

Data Visualization

Because our goal was to turn a set of complex data into a series of patterns that decision makers can easily understand and interpret, we needed an effective visual framework. Our team initially preferred a visualization technique called a "spider diagram," which plotted both magnitude and salience on a circular graph resembling a spider web, with each of the twenty-one impacts (seven each in the biophysical, socioeconomic, and geopolitical systems, as noted earlier) plotted along the outside circumference of the web and the salience values plotted along the radius. However, when we conducted experiments in several large, undergraduate university courses to test the effectiveness of this visualization strategy, we found that most people had trouble interpreting such diagrams. Users with experience in graphic design, abstract reasoning, or mathematics tended to interpret these outputs with greater ease and accuracy, whereas people lacking such a background struggled to make sense of the data conveyed in the diagrams—an unacceptable state of affairs if our goal was to speak to policy makers and other constituent groups without doctoral-level training. Bar charts—with magnitude dictating the height of the bar and salience displayed as color shading—proved to be the most suitable data-visualization technique, one that transcended academic training and cultural background. Model users can input data for magnitude and salience either numerically or by clicking and sliding a set of bars within the computer interface.

Like most large, interdisciplinary research efforts, ours was full of interesting twists and turns. We learned important lessons about how different systems—biophysical, geopolitical, and socioeconomic—are linked together at different scales. A fair amount of our time and effort, particularly during the early phases of the project, was devoted to figuring out how our different scientific disciplines fit together. We found that it was important from the beginning to define key terms such as *system, impact,* and *stakeholders* and to work on forming consensus between project investigators. Most of the researchers on the project agreed that finding a common language—not English or Chinese, but rather an agreed-upon system of meaning for communicating across scientific disciplines—was the most challenging endeavor. Nearly all of the project personnel agreed that the inclusion of "salience" according to the perspectives of multiple stakeholders constituted the most important achievement of IDAM.

Even within the relatively insular world of engineering and hydrological design, researchers are becoming more aware of their own subjective biases and of the ramifications of these biases to the environment and to community members whose lives may be altered forever, but who have little formal power in the decision-making process. By including salience in the model, our intention was to improve transparency in the decision-making processes—which entails not just binary questions such as "to build or not to build," but also questions such as "to build here or to build there" or "to build with these characteristics or those characteristics."

CRITICAL FEEDBACK

The larger question is how such a modeling effort can be applied in the real world to affect the course of policy and decision making. The response to the IDAM among scientists and policymakers thus far has been mixed. Our research group convened a series of workshops to solicit critical input on the model: one at an international conference in the United States that focused on the renewal of the Columbia River Treaty between the United States and Canada; one in Yunnan in which Chinese government officials, hydropower corporation executives, and NGO representatives shared their perspectives with us; and one at the Woodrow Wilson International Center for Scholars in Washington, D.C., to conclude the project and reflect on its potential applications. In each workshop, our goals were to solicit feedback on our approach to measuring dam impacts in a holistic fashion; to test the validity of the specific indicators chosen for the model; and to discuss practical strategies for encouraging policy makers to use this type of tool in their decision-making processes.

The meeting in China, which took place over two days in Kunming, the provincial capital of Yunnan, was particularly instructive because the term *stakeholder* (*liyi xiangguanzhe*) is quite new in China, as is the process of multiperspective decision making in natural resources. Most of the meeting participants were generous enough to commend the effort that went into building the model, stressing the value of a holistic, transparent process for evaluating dams, particularly in China, where the guidelines for integrating environmental and social impacts into the decision-making process are vague at best. One official from a hydropower-development company remarked, "When we think about the effects of a dam, we've got to consider several areas. The first is inundation of land and relocation of people. The second is the effects on the environment."

Mr. Chen, a representative from one of the Five Energy Giants, agreed to attend the Kunming workshop. A middle-aged man dressed impeccably in a business suit, he remained quiet and contemplative throughout much of the day. On the issue of social impacts, however, he spoke quite strongly: "On the socioeconomic side, the relocation problem is key. China is a country with little land and a huge population. It's difficult to give land to resettled people. Cultural protection, especially for minority cultures, is important. It's a national priority [guojia youxian]."

One key concern among the various workshop participants related to the reliability and validity of the data used to run the model. Even under ideal circumstances, these data are compiled from multiple locations or gathered by multiple agencies, which makes standardization a problem. In China, data access can be extremely limited; our team's biophysical group, for example, struggled to get flow data on the Nu River at a finer resolution than monthly averages. Because the Nu is a transboundary river, not to mention a site of international controversy, such data are generally categorized as "for internal use" (nei bu), a euphemism for "classified." The data that are readily available, moreover, are often of suspect quality. The socioeconomic research group, by contrast, was able to construct its own survey to collect the necessary data from households in the two study watersheds. At the end of the day, this sometimes meant that the various research groups were working with data marked by vast differences in quantity and quality.

Moreover, the "Mrs. Wang problem," in which each expert argued that his or her disciplinary area should be represented with more detail or given more weight, came into play on many occasions. Some experts expressed concern over the general structure of the model and how to appropriately assign weights to the various indicators. One incident—indeed, one scientist—came to typify the phenomenon. This particular scientist, Dr. Yang, is a forest ecologist with international training and a great deal of field experience in southwest China. During the stakeholder meetings in Kunming, she spoke passionately about the need to represent the biophysical indicators with more precision and detail, arguing that "just BP2, the indicator for biodiversity, could be broken into a hundred different components." In short, she suggested, the model had sacrificed nuance for simplicity. Dr. Yang's comments pointed to fundamental debates that still exist in the field of conservation biology over how to measure interrelated concepts such as species richness, habitat diversity, and taxonomic uniqueness. Her comments also highlighted the need to think carefully about the trade-offs between simplicity, which makes the tool more accessible to users, and sophistication, which increases its scientific value.

Dr. Zhen, a representative of an international environmental NGO with offices in Yunnan, echoed this sentiment, advocating for more weight on biophysical indicators such as "biodiversity," not only because this indicator related most to his field of expertise, but also because, in his view, the impacts of a dam on habitat diversity in riparian environments may prove to be irreversible. Supporting a precautionary approach, he reflected on the uniqueness and fragility of Yunnan's various ecosystems: "We know that Yunnan is such a small area, but the biodiversity is quite rich. It has as much biodiversity as the entire United States, so we have to pay attention to this unique system. In such a small area, once [biodiversity] is destroyed, you lose millions of years of evolution that you can never get back. . . . Because the biodiversity is so high in a place like Yunnan, it should be rated much higher because if it's destroyed, we'll never even know what was lost."

On the socioeconomic side of the model, IDAM researchers struggled to address the fact that the distribution of a given dam's impacts may be highly uneven and may affect some parties, in particular those with less political power, more than others. One representative from an environmental NGO summed up the challenge by stating, "We're talking about impacts, but it's important to know who bears the impacts." Similarly, a scientist from an academic institution suggested, "We've got to consider the benefit-sharing arrangement, the distribution of benefits [from dams]. This includes doing a stakeholder analysis. Who loses and who wins when it comes to property rights, indigenous knowledge, and so forth?"

Mitigation of the most serious social and ecological effects is a paramount concern for policy makers, one with both scientific and legal dimensions. Adequate mitigation may be possible in certain circumstances but impossible or highly infeasible in others. Mr. Chen, the hydropower company official at the Kunming workshop, suggested, "We've got to consider how able we are to mitigate [jianqing] certain effects of dams, such as ecological impacts. Also, who has the responsibility to mitigate?" This is a particularly difficult challenge in China because institutional accountability varies depending on the size and purpose of a given hydropower project. For example, large dams may arguably have a greater impact on local ecosystems, but national laws require at least a nominal EIA of large-scale projects. Meanwhile, as I have noted, dozens of small "diversion hydropower" projects exist on tributaries of the Lancang and the Nu; because of their small scale, many such projects undergo scrutiny only by county-level officials. Under such limited oversight, mitigation of environmental and social impacts may be compromised (Kibler 2011).[10] Given that the responsibilities for designing, constructing, operating, and regulating hydroelectric dams may fall to literally dozens of agencies and companies, designating responsible parties and holding them accountable for the mitigation of negative impacts become acutely important.[11]

EXPERT KNOWLEDGE FOR WHOM?

In-depth disciplinary knowledge and expertise are necessary but not sufficient to the task of improving the decision-making process for dam development. The Hungarian scholar Michael Polanyi (1962), known for his boundary-crossing insights in fields as disparate as physics and economics, famously envisioned a "Republic of Science" in which highly trained individual experts organize themselves into groups that function collectively as a "body politic" to solve society's most pressing problems. In doing so, these experts must delve deeply into their respective fields, but they must also resurface long enough to fashion a common language that allows them to communicate with one another and to convey their findings to policy makers and to society at large.

In the final analysis, scientific experts—and whatever models or other forms of knowledge they construct—cannot tell us what the best course of action is or even how to optimize the trade-offs between multiple desirable outcomes such as hydropower generation and environmental conservation. These normative questions require a public discussion of the values that drive policy. But where does such scientific input go, and what are its prospects for influencing policy? In chapter 2, I outlined the complex institutional structure of water management in China and alluded to the "nine-dragon" problem in which the roles of various government agencies and private institutions—from the NDRC to the various River-Basin-Management Commissions and the Five Energy Giants that generate electricity—are convoluted and ambiguous. All stakeholders are not created equal.

At the Institute of Water Resources and Hydropower Research in Beijing, I interviewed Dr. Zhou, a junior engineer who was just beginning her career. She had recently led an effort to apply the Hydropower Sustainability Assessment Protocol, developed by the International Hydropower Association, to evaluate the ecological effects of the Jinghong Dam on the Lancang River. Sitting at a large, polished table in a brightly lit conference room at the institute, she reflected on the process of providing technical consulting to Huaneng Corporation, which was responsible for the design and construction of the Jinghong Dam: "Our role is to help them [the hydropower corporations] improve their work. We produce a report based on our findings and offer our opinions about the positive aspects of the project and about aspects that need improving. Sometimes our suggestions relate to the technical aspects of a dam, such as design or operation. Other times, they are administrative suggestions, like 'You should open an office of environmental management to oversee environmental compliance.'"

As the culmination of her research, Dr. Zhou had produced an environmental assessment report (*pinggu baogao*) and delivered it to Huaneng Corporation, but she was unclear on how—or even whether—her efforts would shape the final outcomes of the project. In fact, her experiences with assessment underscore the tenuous nature of environmental law in China: under the current regulations, Huaneng has no statutory obligation to act on the findings of the assessment. It may, if it wishes, simply file away the report into obscurity.

At a more fundamental level, these problems relate to people's ability to participate in the decision-making process, which is ultimately "a question of how much *power* they have to determine the shape and operation of the project" (Nolan 2002:162). In my experience, these crucial dynamics of power and politics generally lie outside the scope of a model. I therefore do not wish to suggest that decision making about dams—whether to build them, how to build them, and how to mitigate their negative impacts—is settled once we have appropriately identified and measured the impacts of greatest concern. Indeed, these steps probably constitute no more than a starting point. The next step, one that is murky and ambiguous but no less important, requires an examination of the participation process itself. Regardless of how well environmental assessments are designed and carried out, if such scientific input does not fit into a larger system of equitable, transparent, and accountable decision making, it is of little value. Chapter 6 addresses key social challenges such as public participation and the implementation of a rights-based framework in decision making about dams, particularly when people's livelihoods are at stake.

6 |

PEOPLE IN THE WAY

Resettlement in Policy and Practice

DURING AN interview at the Institute of Water Resources and Hydropower Research in Beijing, Dr. Yin, a senior engineer and colleague of Dr. Zhou (the junior engineer mentioned in chapter 5), reflected on the key challenges that his organization faces as it seeks to implement what it calls "sustainable-hydropower development" (*shuidian kechixu fazhan*), a concept that is now featured prominently in the glossy brochures that advertise the organization's mission. Dr. Yin remarked,

> Sustainable-hydropower development has two dimensions: the environmental dimension, which can usually be addressed with technical solutions; and the social dimension, which has to be addressed through policy. Our primary focus is on the engineering and technical aspects. For example, how can we better understand and evaluate the effects of hydropower development on river ecosystems? We need better evaluation methods. It's very challenging. But the technical problems are relatively easy to solve, compared with cultural and social problems.

Dr. Yin's acknowledgment of the social costs of dams may seem like a fairly enlightened position for an engineer, but in my experience it is not unique. In many interviews and public interactions, even the most technically inclined hydropower experts tended to recognize that their industry could not be successful over the long term in the absence of a sound legal, policy, and institutional framework for dealing with the social consequences of dams.

Although this goal is widely shared, the means of accomplishing it are not so clear-cut. By a wide margin, the most pressing social problem associated with dams is the uprooting of communities through displacement. Riparian, or streamside, environments happen to be both the prerequisite for dam construction and a common place in which rural people live and work. Displacement caused by dam construction and reservoir inundation can be seen as the first-order social consequence of dam building, which then causes a cascade of second- and third-order social ills: the disintegration of community identities and networks; health problems; unemployment; and sometimes overt social conflict (Scudder 2005; Cernea 2000).

The twentieth century and the sliver of the twenty-first that we have seen so far have been marked by the migration of hundreds of millions of people. In many cases, their movement has involved at least some volition and self-determination: people move to avoid environmental disasters, to seek greater political freedom, or to maximize their family's economic chances. But people also move because they are forced or otherwise coerced by factors beyond their control. Resettlement to make way for development projects—infrastructural projects such as hydropower facilities, natural-resource extraction, tourism development, conservation areas, and urban expansion requiring the annexation of rural land—has displaced many millions of people against their will. These "people in the way," as the anthropologist Anthony Oliver-Smith (2010) calls them, are often nameless and faceless, collectively adding up to a number that is probably impossible to calculate with any precision. The WCD (2000) can only say that between 40 million and 80 million people around the world have been displaced by dams, a fairly wide window of uncertainty. For people who undergo involuntary displacement and resettlement, who experience the uprooting of their families and communities, such struggles often represent "disasters of development" (Oliver-Smith 2010:1).

In 2004, Xinhua News Agency, the CCP's official media service, released a short report based on research conducted by the MWR, which concluded that the nation's displaced population related to dam construction totaled at least 15 million people, ranking it first in the world (Yao 2004). The tone of the report is strangely ambivalent. On the one hand, it cites studies by the World Bank and the Asian Development Bank that conclude that most of China's dam migrants relocated willingly, a claim that seems dubious given the poor standards of informed consent and public participation in such projects, particularly before the 1990s. On the other hand, the report provides a frank assessment of the myriad shortcomings of past and present resettlement policies—poor planning, insufficient levels of compensation, and the selection of unsuitable relocation sites—and alludes to broad public dissatisfaction and even social unrest related to dam-induced displacement.

The vast majority of studies on the social impacts of displacement and resettlement conclude that lives and livelihoods are never the same afterward. In *The Future of Large Dams*, Thayer Scudder (2005), an anthropologist who participated in the WCD's comprehensive review process, concluded after perhaps the most extensive global review of large-scale dam development ever undertaken that in 80 percent of cases—from Africa to Latin America to Southeast Asia—resettled people experience a marked decline in their standard of living. But understanding the effects of resettlement on rural communities in China requires a close look at the details of policy governing resettlement for hydropower projects and at the ways individuals participate in the decision-making process. It also requires an examination of the changing nature of land-use rights in contemporary China, one of the most problematic and conflict-riven issues of the reform era. The story involves social scientists who work in the growing field of social impact assessment and seek to improve current resettlement practices or, at a minimum, to mitigate some of the worst outcomes for vulnerable people.

SOCIAL IMPACT ASSESSMENT

In collaboration with many colleagues, I have been working to refine a framework for assessing the social effects of dams based on internationally accepted standards in the growing field of social impact assessment (SIA). As an anthropologist, I tend to view dams as emblems of both geophysical and social engineering and to ask how we can better understand and mitigate the impacts of dams on communities, economies, and cultures. The present framework for conducting SIA in China is extremely limited. If a project is funded by an international financial institution such as UNDP, the World Bank, or the Asian Development Bank, these organizations typically expect their own guidelines to be followed as a condition under which financing is allocated. As I pointed out in chapter 5, the new EIA Law provides a legal basis for SIA and public participation. Article 5 of the law reads: "The government encourages relevant entities, experts, and the general public to participate in appropriate ways in the environmental impact assessment process" (Chinese National People's Congress 2002a). But there is currently little formal structure guiding practitioners about how such a process would work in practice.

The difficulties of assessing the social costs of dams serve as a reminder that the key issues at stake transcend the merely technical. Beyond the physical and ecological impacts associated with hydropower projects, political leaders and citizens alike are concerned about the geographical distribution of electrical power and water resources, the administrative decision-making process, the inclusion of relevant stakeholders, the relocation and resettlement of displaced people, and the disruption of social, cultural, and economic life in communities affected by dam construction. SIA is geared toward understanding these outcomes at a point early enough in the development process to make a positive difference. It can be defined as "the process of analyzing (predicting, evaluating and reflecting) and managing the intended and unintended consequences on the human environment of planned interventions (policies, programs, plans, projects) and any social change processes invoked by these interventions so as to bring about a more sustainable and equitable biophysical and human environment" (Vanclay 2002b:388).

The goals of SIA are quite straightforward. By identifying potential impacts in advance of a large project, government agencies and private development interests can make better decisions about which dam sites and design specifications are likely to cause what kinds of social outcomes. They can also begin to plan at the outset of a project how to best mitigate the worst social impacts and how to properly compensate affected individuals and communities. These goals are part of a larger trend in international development over the past several decades that affirms the belief that "human beings are at the center of concerns for sustainable development" (UNEP 1992:Principle 1).

The United States was the birthplace of SIA, and most SIA in the U.S. context takes place under a federal mandate. Since the passage of NEPA in 1969, EIA has become an integral part of the environmental decision-making process in the United States. Under NEPA, federal agencies must file comprehensive environmental impact statements prior to undertaking any actions with the potential to significantly affect the quality of the human environment. These statements typically include a social science component that assists agencies in understanding the social consequences of policies, programs, and projects. In 1994, the Interorganizational Committee on Guidelines and Principles for Social Impact Assessment (2003, 1994) produced basic guidelines for conducting SIA in federal projects, updating the guidelines again in 2003.

SIA does have some fundamental biases and assumptions that come from Western philosophical and legal traditions and that further complicate the implementation of SIA in international contexts. Because social, cultural, and political conditions differ in disparate locations, conducting SIA in international contexts can be particularly challenging. Responding to the need for internationally relevant guidelines for SIA, the International Association for Impact Assessment developed a set of principles that are broadly applicable to large development projects. These principles include, among other things, a dedication to the precautionary principle, a balancing of intragenerational and intergenerational equity, the preservation of social and cultural diversity, and the internalization of costs associated with a planned project (Vanclay 2003).

As we have seen in the case studies on the Lancang and Nu Rivers, resettlement causes a wide array of subsequent social impacts, including changes in household size and structure; changes in employment and income-generating opportunities; alteration of access to and use of land and water resources; changes in social networks and community integration; changes in the nature and magnitude of various health risks; and even a disruption of displaced individuals' psychosocial well-being. Managing and mitigating these impacts are important tasks because, as the WCD noted in its seminal report, these effects are "spatially significant, locally disruptive, lasting, and often irreversible" (WCD 2000b:102).

One of the key challenges of assessing the social impacts of dam projects is establishing a standard set of variables to measure. As Frank Vanclay has noted, "The variables that are important must be locally defined, and there may be local considerations that a generic listing does not adequately represent" (2002a:200). The process of SIA itself provides a partial solution to this problem, however, since the step-by-step process involves an in situ evaluation of stakeholder identification and scoping of activities likely to result in impacts. Although the important variables may differ considerably from project to project, a comprehensive SIA process should allow practitioners to identify and measure locally salient variables. Table 6.1 shows the various steps involved in using SIA for dam projects.

TABLE 6.1 Steps in the Social Impact Assessment Process for Dam Projects

Step	Description	Importance/Significance
Step 1	Identify interested and affected individuals and communities (stakeholders).	Failure to include all stakeholders can result in improper assessment of impacts. For dam projects, stakeholders may include relocated people, upstream and downstream residents, communities affected by roads and transmission lines, and conservation groups concerned about environmental impacts.
Step 2	Facilitate the participation of stakeholders in the decision-making process.	Facilitating participation ensures that all affected individuals are included from the beginning, which increases the likelihood of local support for the intervention, minimizes impacts, and begins the process of considering measures to mitigate or compensate. All stakeholders should be able to contribute to the selection of variables to be considered in the SIA.
Step 3	Collect baseline data (social profiling).	Data sources may include scientific literature, census bureaus or other agencies, or primary research such as surveys, interviews, and so on. Both qualitative and quantitative research methods may be used. Data collection ensures that demographic, economic, health, social, and cultural information is understood about the present state of the community before the intervention, thus providing a baseline for comparison after project completion.
Step 4	Identify and describe the activities that are likely to cause impacts (scoping).	Impact-causing activities should be described in enough detail to help identify what data are needed to predict impacts. For example, practitioners should assess the footprint of the reservoir, timeline for construction, number of people to be displaced, and other key variables.

TABLE 6.1 (Continued)

Step	Description	Importance/Significance
Step 5	Predict likely impacts and determine how stakeholders may respond.	Prediction compares the present baseline conditions with likely conditions following the intervention. Direct impacts (such as relocation) and secondary impacts (such as change in employment status, etc.) must be considered in sufficient detail to allow monitors to judge when postresettlement living-standard goals have been met.
Step 6	Identify possible intervention alternatives (including a nonintervention alternative).	Identification provides an array of alternatives for the location and design of dam projects. Each alternative should be assessed separately so that decision makers can choose one that is both technically and financially feasible and minimizes environmental and social impacts.
Step 7	Recommend mitigation or compensation measures.	Mitigation or compensation measures may be built into the selected intervention alternative. Practitioners should also identify the agency or organization responsible for mitigation or compensation.
Step 8	Develop monitoring and management programs.	Development of such programs assures that impacts are managed through the four phases in the life cycle of a dam, including planning, construction, operation, and decommissioning. It allows practitioners to compare actual impacts with projected impacts.

Source: Adapted from Tilt, Braun, and He 2009.

The legal framework for SIA in China has been slow to take hold but is gaining momentum. In the 1990s, the Institute of Investment Research within the State Planning Commission (the predecessor of the NDRC) issued guidelines and methods for SIA in its large investment projects (Li and Shi 2011; NDRC 2007). At the eighteenth National People's Congress in 2012, which ushered Xi Jinping into the top leadership position, the State Council passed a new directive on "social risk assessment" (*shehui fengxian pinggu*) that provided a stronger administrative mandate for considering the social costs of large development projects before they are initiated. In an interview with the *New York Times*, Environmental Minister Zhou Shengxian reflected on the importance of the directive, saying, "No major projects can be launched without social risk evaluations. . . . By doing so, I hope we can reduce the number of mass incidents in the future" (Bradsher 2012). *Mass incident (quntixing shijian)* is a euphemism commonly used to refer to public protests and other forms of dissent. It is too early to tell what kind of statutory authority the "social risk assessment" directive will have—whether it will eventually be legislated into law or passed as a State Council ordinance—but it signals unequivocally the political leadership's recognition that the mounting social costs of development projects need to be addressed at high administrative levels.

Assuming that the institutional and legal challenges of SIA can be addressed, which is admittedly not a sure bet, there are even more basic scientific questions that confront SIA practitioners in China and elsewhere. One of the most problematic aspects of conducting SIA of large dam projects is ensuring that the analysis takes place at the proper temporal scale. For example, multilateral development agencies, national governments, and private developers alike tend to monitor primarily the initial years of resettlement, which has the effect of missing the long-term problems that communities must face. Evaluations during the initial years of resettlement may give a false reading of success, effectively sidestepping the question of how social impacts unfold over time. In the case of dam projects, practitioners should ideally think about a project's full life cycle and anticipate impacts at each of four stages (Sadler, Vanclay, and Verocai 2000), including conceptualization and planning, construction, operation, and decommissioning (Interorganizational Committee on Guidelines and Principles 1994)—something that is rarely done.

As the discussion of Lancang dams in chapter 3 highlighted, people whose lives were uprooted by the construction of Manwan Dam in the

1990s had a difficult time obtaining adequate housing, finding employment, and securing a decent standard of living for their families. The problems have now persisted for a generation. Many experts and government officials have acknowledged that resettlement plans at the Manwan Dam site were carried out haphazardly and that compensation levels were woefully inadequate.

In addition to Manwan, the socioeconomic survey results provide some insight into how resettlement is affecting villagers at three other dam sites on the Lancang at various stages of construction. Resettlement has radically changed households' access to farmland and other means of subsistence and the way they relate to their neighbors and kin through social networks of reciprocity. In many cases, resettlement has provided a powerful incentive for households to send one or more family members into cities and towns to engage in wage labor and send remittances home. These findings—with all their deficiencies and framed within a window of time of less than two decades—nevertheless represent some of the best knowledge to date about how dams affect people. For the thousands of other projects throughout China, the answer is that we unfortunately don't know that much.

Analyzing the correct spatial scale can be equally troublesome. At what geographic location or level of analysis are the effects of dam projects best examined or understood? While people living near a dam site or reservoir may experience drastic negative impacts, the net effect downstream may be positive due to increased reliability of irrigation water or the benefits of flood protection. Furthermore, if we consider impacts from a regional or national scale, large dam projects may appear to offer a net benefit due to increased hydropower or revenue from the sale of electricity. Dams in southwest China provide a growing share of electricity to fuel rapid industrial development in coastal areas with access to global capital some 2,000 kilometers away (Magee 2006)[1]. In the case of large dam construction on transboundary rivers, the situation becomes even more complex due to limited data availability and geopolitical considerations. On the Lancang River, for example, downstream nations including Laos, Myanmar, Thailand, Cambodia, and Vietnam have experienced biophysical, ecological, and socioeconomic impacts from China's decision to seriously alter the hydrograph of the upper Mekong River, an issue that will be examined more closely in the next chapter.

Certain socioeconomic variables, such as income or housing values, are easier to identify and measure than others. Mitigation policies often fail to recognize sociocultural impacts and underestimate the economic and social value of prior livelihood strategies. Many assets, in particular those that are communally owned and managed or are nonmaterial, are not likely to be prioritized in remuneration plans and may not be compensated for at all. Irreplaceable losses—cultural connections to ancestral lands, for example—are impossible to evaluate in a cost–benefit analysis and consequently are often externalized by development authorities (WCD 2000b). On Yunnan's rivers, displacement of minority nationalities raises difficult questions about how to assess nonmaterial losses such as community ties, cultural heritage, and traditional ecological knowledge.

Dr. Li, a sociologist who worked for a government-sponsored research institute, had more than a decade of experience conducting SIAs for hydropower projects. He discussed with me some of the methodological challenges of applying the SIA approach in China. His research team, charged with the task of collecting baseline data for a poverty-alleviation project in rural Guizhou, adopted a standardized household survey used by Western colleagues, translating it into Chinese and modifying it only slightly. One of the questions asked study participants to indicate their occupation and contained a range of possible responses including "farmer," "wage laborer," "government official," and so forth. The research team struggled to fit their data into these predetermined categories, a process that became truly ridiculous when one study participant identified herself as a matchmaker (*meipo*), whose job was to use astrological charts to determine the suitability of potential marriage partners for local villagers. Recounting the story, Dr. Li was nonplussed; he chuckled, "How were we supposed to fill out the questionnaire for that?" Sometimes the epistemological categories that we are expected to apply to a topic of study prove inadequate to the task.

THE ROLE OF EXPERTS

As part of my work to understand the scientific and policy dimensions of hydropower, I interviewed many social scientists like Dr. Li who had conducted SIA work for water-resource projects in recent years. I also reviewed the growing body of published literature in English and Chinese on the subject. In the process, I gained an understanding of the challenges of resettlement planning and implementation, the mechanisms through which social scientists try to improve the outcomes, and the strategies used by displaced individuals and communities to get what they need. Reviewing the recent history of SIAs, even those conducted at the behest of government agencies, one doesn't have to wait long before encountering the social problems—poverty, mass protests, and even violence—that go along with large-scale development projects.

The story of Dr. Ying Xing, a professor at the Chinese Academy of Social Sciences, provides a point of entry into this topic. As a doctoral student in sociology in the 1990s, Ying was invited by the Three Gorges Project Construction Committee to study the resettlement process in the Yangtze River region and provide critical feedback on how the committee could better serve the needs of affected communities. His report, entitled *The Story of Dahe River Migrants' Petition: From "Seeking an Explanation" to "Rationalization"* (2005), was translated into English by Probe International, an environmental advocacy group based in Canada. It has quickly become a touchstone for controversy between Chinese government officials and advocacy groups concerned about the rights of displaced people.

Ying was given the honorary title of vice governor of Yunyang County in Chongqing Municipality, presumably to expedite his access to local-government officials who might have otherwise obstructed his way. Although his position carried no official authority, it afforded him a fair amount of latitude to go about his work in the hopes that he could make some recommendations to improve resettlement for the Three Gorges Project. Scouring the county archives and interviewing local residents, Ying discovered that the social problems related to resettlement in the 1990s were only the beginning and were in fact compounded by "leftover problems" (*yiliu wenti*) from a previous resettlement process that had taken place in the 1970s to make way for the Dahe Dam on a tributary of the Yangtze.

Ying discovered a record of villagers' complaints and protests that spanned nearly two decades, from 1977 to 1994; his careful research provides a kind of archaeological glimpse into what can go wrong during the resettlement process, particularly in the absence of political accountability. Villagers throughout Yunyang County lodged a litany of complaints: underreporting of land-area measurements during the requisition process; undervaluation of housing for compensation; and failure to account for inflation, which made it impossible for some villagers to build homes of similar size and quality to the ones that were inundated by the reservoir of Dahe Dam. One of Ying's study participants, with the pseudonym "Mr. Liu," commented on the inadequacy of compensation measures during a retrospective interview, suggesting that the government was ultimately culpable: "If you break my bowl, you should compensate me with a bowl of the same size and quality; if you smash one of my jars, you should replace it with a similar one" (qtd. in Ying 2005:3).

A common complaint underlying villagers' protests was the obvious discrepancy between central-government plans and the actions of local-government functionaries tasked with carrying out the plans. Many villagers saw a system that was rife with corruption, though Ying was more balanced in his assessment. He found little evidence of overt corruption but did note that government officials may have siphoned off project money that had been earmarked for the compensation of villagers. Ying concluded that they had done so not for personal gain but to allocate funding for local projects such as roads and schools. Nevertheless, in the eyes of many villagers, such tactics simply depleted the amount of compensation that found its way into their own pockets (Ying 2005:6).

In the Dahe Dam case, the scope of social problems associated with resettlement extended far beyond the individuals and families who were forced to move. One group of farmers adjacent to Dahe Dam was targeted for relocation but could not move because the community where they were to be resettled had not been properly contacted and consulted. Villagers in this new "host community," anticipating the loss or reallocation of their own farmland plots, put up fierce resistance. There were also difficulties in determining the proper level of compensation because several households had had children in excess of the Planned-Births Program in the 1980s, and these children were erroneously given farmland allocations under the Household Responsibility System. Would compensation for lost land extend to them?

The heart of Ying's report was an examination of the strategies used by villagers to seek redress from the social ills of displacement. When the Dahe Dam was constructed in the 1970s, the nation was in the grip of the Cultural Revolution; concepts such as individual rights and rule of law, tenuous even at the best of times, were at their lowest ebb.

Nevertheless, villagers found creative strategies to voice their concerns. In Ying's words,

> Especially before the mid-1980s, few Chinese citizens took legal action, or were even able to access appropriate legal weapons, in the face of economic disputes or social injustice. Instead, many turned to party or government organs for help, through "looking for an upright official" [zhao qingtian] or "seeking an explanation" [tao shuofa]. The use of "petitioning" [shangfang] became a frequent and widespread legal practice unique to China. Citizens were not demanding universal rights, but rather, were seeking the satisfaction of specific, personal interests. They often sought the help of officials at various levels in the belief that the party and government sincerely wanted to take care of ordinary people.
>
> (2005:10)

In Yunyang County—in particular Shanyang Township, one of the communities most affected by the Dahe Dam—acts of protest unfolded over many years and involved public officials from Beijing to the provincial government and even down to the county, townships, villages, and production cooperatives. Success is not easily measured in cases like these. According to Ying, villagers managed to get compensation money to move down the bureaucratic ladder from the county to the township to the production cooperatives (xiao zu). They also succeeded in lobbying officials to lower their grain quotas and to inject higher government subsidies into a local calcium carbide factory where many displaced people worked after losing their land.

Petitioning (shangfang), which has been a mainstay of citizen–state relations in China for a long time, was the tactic that proved most effective in the Dahe case. Ying created a typology of shangfang that describes the conditions under which this tactic is most likely to meet with success. Collective action usually proves more effective than individual action, provided that the scale is kept modest enough to avoid direct threats to governmental power. Leapfrogging over unsympathetic, midlevel officials in order to appeal at the top of the political hierarchy can be effective, but this tactic often worries the central government since it poses a risk for broader social conflict. And timing is crucial: sensitive political moments such as the lead-up to the Beijing Olympic Games or the transition of political leadership can precipitate a crackdown (Ying 2005).

In practice, *shangfang* can involve a variety of tactics, from letter-writing campaigns to signed petitions to amassing complainants in front of government offices in an attempt to gain public support. The *shangfang* process can clearly serve as a mechanism for extracting accountability from political elites, but it also tends to unfold within well-defined cultural boundaries.[2] Villagers tend to be most successful when their requests are seen as modest and reasonable; otherwise, they risk being labeled as belonging to "nail households" (*dingzi hu*) that chronically seek to stir up trouble. In this regard, *shangfang* is not unlike the "weapons of the weak" described by James Scott: "The antagonists in such contests . . . know each other's repertoire of practical action and discursive moves. There is, in other words, a kind of larger social contract that gives some order and limits to the conflict. . . . The limits and constraints characterizing conflict are never cut-and-dried to the participants. The antagonists are, all of them, continually prospecting new terrain—trying out new stratagems and wrinkles that threaten to change, and often do change, the shape of the 'game' itself" (2005:398).

Ying's experience unraveling the complex social problems related to dam-induced displacement is shared by many other social scientists. Dr. Li, for example, the sociologist who worked for a government-sponsored research institute, had recently conducted an assessment of the long-term social impacts of resettlement for the Three Gorges Project, for which at least 1.3 million people have been relocated over the past twenty years. In a teahouse near his office in Beijing, when I asked him whether he could provide me with a copy of his report to the Three Gorges Commission, he replied flatly, "No, the report is for internal use [*neibu*] only." He did, however, outline the main findings of the study, reciting a list of ongoing problems that were by now all too familiar: "First, as we've already discussed, compensation levels were too low. Employment was also a problem; job-training programs were not always very effective. There isn't nearly enough farmland in the region to support all the displaced farmers. In some cases, farming households were widely dispersed; in other cases, they were given adjacent pieces of land almost like collective farms from the old days. The only difference is that no one is in charge; there is no production team leader."

Dr. Li also shared with me, in general terms, the recommendations that his research team made to the Three Gorges Commission, the most important of which was that the central government should go beyond household-level compensation to provide funding for ongoing community-development projects that foster social cohesion. In light of the serious social problems that his team and others have documented for the Three Gorges Project—including conflict between migrants and members of host communities as well as a dramatic demographic shift in which most young people have left their villages in search of wage labor—he argued that the central government must play a stronger role. "When we talk about resettlement," he concluded, "we have to emphasize three things: rights [*quanli*], responsibilities [*zeren*], and benefits [*liyi*]. They are all important."

PUBLIC PARTICIPATION

If SIA is to become an effective tool in China, one that provides meaningful and timely input into the decision-making process regarding hydropower development, its practitioners will need to remain faithful to core principles such as public participation. But they will also need to exercise flexibility and creativity in order to operate within the "unique historical, structural, cultural and practical barriers to participatory planning" in contemporary China (Tang and Zhan 2008:58).

This brings us to a key point about the role of public participation in development initiatives. "Participatory development"—an approach that makes people "central to development by encouraging beneficiary involvement in interventions that affect them and over which they previously had limited control and influence" (Cooke and Kothari 2001:5)—has been a principal aim of multilateral development agencies since the early 1980s. UNDP acknowledges the importance of public participation in its Millennium Development Goals, as does the World Bank, which has published its own guidelines for carrying out participatory development programs. The latest term for this practice in the World Bank is *community-driven development*, which regards people as "assets and partners in the development process" and gives "control of decisions and resources to community groups and local governments" (World Bank 2005).

Over the past two decades, participatory development has evolved in at least two fundamental and significant ways. First, the idea has moved from a set of concrete steps to a more general, abstract attitude regarding the need to engage stakeholders and build social capacity. Indeed, many development agencies see the promotion of "social capital," "civil society," or even "empowerment"—all of which were once seen as stepping stones toward more concrete development outcomes such as improved nutritional status or more sustainable land-use practices—as laudable goals in their own rights. Second, the idea of participatory development has been scaled up: where the "community" was once the important level of analysis, broader participation in governance and institutional decision making are now seen as tenets of the participatory framework.

In regards to employing a participatory framework for dams, different constituencies may have radically divergent ideas about whether and how a given project should proceed; in such cases, the key question becomes how to ensure that these various voices are heard. The landmark report published by the WCD in 2000 emphasizes the application of what it calls a "rights and risks" approach to evaluating large dam projects. This approach includes, among other things, "self-determination and the right to consultation in matters that affect people's lives, the right to democratic representation of people's views on such matters, the right to an adequate standard of living, [and] freedom from arbitrary deprivation of property" (WCD 2000a:200).

This approach draws upon fundamental human rights frameworks agreed upon by the international community, including the UN Declaration of Human Rights (1947), the Declaration on the Right to Development (1986), and the Rio Declaration on Environment and Development Principles (1992) (see UNEP 1992; UN General Assembly 1986, 1947). From a strictly pragmatic standpoint, the encouragement of stakeholder participation in the decision-making process increases the likelihood that dam projects are economically viable, socially equitable, and environmentally sustainable (World Bank 2005; WCD 2000a:202). In my conversations with Dr. Li, he was adamant that broad public participation, grounded in well-defined rights, was a fundamental requirement of just and sustainable-development projects. He remarked, "Only with rights can there be consultation; only with consultation can we find a way forward [*You quanli, cai you xieshang; you xieshang, cai you banfa*]."

Despite the indisputably top-down nature of Chinese politics, the ideal of participatory development is gaining traction in China, at least within official rhetoric. Pan Yue, the famously outspoken environmental protection vice minister, has even gone so far as to suggest that the cadre evaluation system (*kaohe zhidu*), which has long been the primary mechanism through which higher levels of government control the career trajectories of subordinate officials, include one criterion that would judge how effectively local officials carry out public participation: "First of all, we must understand clearly that public participation is the right and interest of the people endowed by law. . . . Involving public participation in environmental protection should be an aspect by which to evaluate political performance" (Pan 2006).

Depending on how "participation" is structured and implemented, however, it may not be an unmitigated good; as social scientists working in the international development arena have pointed out, democratic principles and individual rights are not cultural universals, which means that powerful constituent groups can often dominate the participatory process in order to shape project outcomes in their favor (Mosse 2005; Cooke and Kothari 2001).

Ironically, Chinese officials thus far seem to have interpreted the mandate for public participation largely as a call for "expert participation." Professionals from academic institutions or government ministries are often called upon to provide testimony at public hearings for large development projects, while the "old hundred names" (*laobaixing*)—the common people—are poorly represented. Dr. Li and one of his colleagues showed me several social media sites on the Internet where people roundly criticized the role of expertise in environmental management and in governance issues more generally. One Internet meme in particular has enjoyed wide circulation: people refer to experts (*zhuanjia*) by adding a "stone" radical to the left side of the first character, a change that retains the sound but changes the meaning of the term from "expert" to "brick house." The result is a fairly unsubtle critique of experts—who claim to know what is best, who make policy recommendations, but who are not held accountable for the consequences and who are ultimately used as tools by those with political and economic power.

The "rule of experts" is especially evident in the higher echelons of political leadership. As of 2000, more than half of all top political posts in China were filled by CCP members with engineering or technical college degrees. This trend holds for the hydropower sector, too: former premier Li Peng, for example, who trained in hydroelectric engineering at the Moscow Power Institute in the 1940s and eventually served as minister of electrical power, became one of the most strident proponents of the Three Gorges Project (Yeh and Lewis 2004:454).

Li Peng's son, Li Xiaopeng, became the first head of China Huaneng Group, the largest of the Five Energy Giants that were spun off from the Ministry of Electrical Power in 2002 and the one that holds hydropower-development rights on the Lancang River.

RIGHTS AND REDRESS

The rise of petitioning (*shangfang*) as a primary strategy for addressing the problems of displacement can be explained in part by the lack of other viable alternatives. Access to the judicial system, like many aspects of political life in contemporary China, is a difficult pursuit, for a variety of reasons. First, only individuals directly harmed by a given project have the legal right to sue. This means that although domestic and international environmental organizations can provide moral, logistical, and even financial support, they cannot be plaintiffs in a lawsuit. In 2012, the Civil Procedure Law was amended to allow "government departments and concerned organizations as designated by law" to engage in public-interest litigation, and this amendment will likely result in a dramatic increase in the volume of environmental lawsuits in the near future (Ngo 2012; see also Wang 2007).

Second, in the case of collective lawsuits involving many parties, each individual party must opt into the lawsuit and provide a copy of his or her national identification card (*shenfen zheng*), which can make collective lawsuits logistically complicated and politically risky because plaintiffs may fear reprisal. Furthermore, Chinese courts are notoriously difficult to access for those who lack political clout. Although lower-level courts are easier for plaintiffs to access, they are particularly vulnerable to the political influence of big players such as state-owned firms or private and shareholder corporations. As the legal scholar Rachel Stern concludes following an analysis of environmental litigation related to pollution damages, "Despite occasional successes, civil environmental litigation remains a weak tool for environmental protection" (2011:310). For all its shortcomings, *shangfang* remains one of the only readily accessible tactics for many people seeking redress.

Perhaps the most intractable problem standing in the way of a rights-based approach to resettlement policy in China relates to the changing political economy of rural land. During the Maoist period, agricultural production was controlled by the state via a network of communes, production brigades, and production teams overseen by party cadres. This system became the basis for economic and social life in rural China for nearly three decades. The Household Responsibility System, introduced in the 1980s, granted farming households the right to make their own production decisions and earn residual income from their land by selling crops on the market, which was subject to fewer and fewer government restrictions. This effectively reestablished China as the world's largest smallholder farming system.

But the property-rights regime governing access to rural land remains one of the greatest headaches for political officials and villagers alike. Property rights are ultimately grounded in socially recognized claims, whether through custom and tradition or through formal, codified law. Far from homogenous, property rights may encompass a wide variety of entitlements, from resource access and use to formal ownership and transferability to a third party (Ribot and Peluso 2003). In China, the rural land-rights regime exhibits considerable variability in different regions but is commonly seen to contain four fundamental dimensions: the right to residual income generated from agricultural activity on land; the right to use land in relative freedom from state regulation and other encumbrances; security of tenure rights into the future; and land-transfer rights (Liu, Carter, and Yao 1998).

The Household Responsibility System, arguably the backbone of the reform-era economy, gives farming households access to cash income through agricultural sales. Decisions regarding crop selection, cultivation, and market distribution are made at the household level, and economic risk within China's rapidly changing market economy is also assumed at the household level. Both security of future tenure rights and transferability to a third party are major issues of concern in rural China. In contrast to urban land, which is owned by the central government (*quanmin suoyou zhi*), rural land rights are vested in rural collectives (*jiti suoyou zhi*) at the level of township, village, or production cooperative—an arrangement that is codified in the PRC Constitution. Individual farming households are typically granted certificates that give them use rights, along with the right to sublet land within the term of their lease, but not full ownership rights over two types of land leased from the rural collective: "responsibility land" (*zeren tian*) and "contract land" (*chengbao tian*). The length of rural land leases has increased dramatically since the 1980s and now spans seventy years for responsibility land, although the collective retains the right to appropriate land within its jurisdiction when necessary. On contract land leased from their collectives, farmers now enjoy "indefinite" lease periods; government planners hope that this relative sense of long-term security will help to promote specialization, increased land transfers, and economies of scale in the agricultural sector.

Villagers face other encumbrances on their land rights, including limited knowledge of how the complex and ever-changing land-tenure system works. Rural land cannot be used for collateral on a loan, nor can it be sold for nonagricultural purposes; instead, it must first be converted to state-owned land and then sold, a process over which the local government retains monopoly control. Local-government officials, whose performance targets are set by higher administrative levels, are thus incentivized to "chase commerce and attract investment" (*zhao shang, yin zi*) by pursuing land-conversion strategies that are most remunerative (Whiting 2011).[3]

This shift in agricultural governance has greatly increased farming income, a welcome trend for most rural people. But it also creates considerable risk for individual farming families, who must meet their own economic needs as the state provides less security and fewer services than during the socialist period. The implications for dam-related resettlement are clear. If property rights exist on a spectrum—from informal access rights on one end to the right to earn a living by farming in the middle and full ownership and salability on the other end—the PRC effectively recognizes only the middle part of the spectrum. When villagers are forced to move in order to make way for a development project, the law requires that they be compensated, but only for lost income from agricultural production, not for lost ownership of the land and whatever future value or productive capacity might be embodied in it. The law does not recognize ownership or salability as part of the land-rights "bundle" belonging to rural people. Nor does it recognize, at the other end of the spectrum, informal resource-harvesting rights on land that is communally managed.

To provide just one example of how the land-rights regime affects displaced people, villagers in the Nu River Gorge have historically relied on an incredibly wide assemblage of resources over which they hold no formal title at all. Many households still gather mushrooms, herbs, and firewood on forest land under informal agreements with their relatives and neighbors, consuming these products as part of their daily subsistence and sometimes selling them on the local market to supplement their income from farming. Villagers also grow corn on recently cleared hillside land that probably does not belong to them at all; in the socioeconomic survey results described in chapter 4, it was not uncommon for villagers to report planting more than ten *mu* of corn as a cash crop while reporting total land holdings that amounted to only half that figure. Many forms of land use in rural China are, strictly speaking, extralegal, but they are nevertheless extremely common. Loss of this land may not fit the definition of "dispossession" but may in fact be something worse: the requisitioning of assets that vulnerable people rely upon but over which they have never held security of title.

As the economist Hernando de Soto (2003) has observed, the single greatest obstacle faced by most of the world's poor is a lack of formal property rights. One need not be a proponent of private property to see that clearer and more transparent property rights are the basis of a stable economic system. Well-defined property rights of whatever type—whether vested in the individual, the community, or the state—are the ultimate source of wealth, especially in rural areas where land constitutes people's most productive asset and their most abiding source of future security. Do powerful interests such as government agencies or hydropower corporations actively seek to undermine the livelihoods of the poor through a process of dispossession? Perhaps. But they don't really need to because the existing, ambiguous property-rights regime has already accomplished that for them.

PEOPLE IN THE WAY

The dismantling of the State Electric Power Corporation in 2002 and the distribution of its assets to various state-owned enterprises and shareholder corporations constituted an application of market logic to China's electric-power sector. As I have already described in detail, this distribution set in motion a process in which the Five Energy Giants and their various subsidiaries have come to shoulder many market-driven imperatives: they are expected to operate efficiently and profitably in order to generate value for investors. In a fundamental sense, displacement and resettlement to make way for the dams in Yunnan represent villagers' head-on encounter with the market logic of reform-era China.

Proponents of liberal economic policy tend to attach a range of salutary adjectives to the market. Whether referring to a specific institution such as a locally owned store or a Wall Street firm or to a more abstract, idealized concept, the market is often described as "efficient," "rational," "competitive," and even "free." But if the market is the only measure of something's value, then only things readily measured in market terms will show up on the balance sheets. Nonmaterial assets—the substance of human relationships or the long-standing attachment to a place and its resources—tend not to be accounted for in any meaningful way. In Marxist political economy, such uncalculated costs are often referred to as the *faux frais* of production, the "incidental operating expenses" that accompany any business transaction but do not add value to the exchange and are therefore willfully overlooked. Marx argued in *Capital* that dispossession and displacement were not incidental to the economic development process, but rather central to it: "Along with the surplus population [of unemployed workers], pauperism forms a condition of capitalist production, and of the capitalist development of wealth. It enters into the *faux frais* of capitalist production; but capital knows how to throw these, for the most part, from its own shoulders on to those of the working-class and the lower middle class" ([1867] 1984:602–603)

In the process of building a dam, there are various expenditures—fixed capital for infrastructure, labor costs, and operating costs, to name only a few—that hydropower-development companies and their financial backers justifiably consider productive investments for which they can expect a monetary return when electricity is sold and distributed on the grid. For example, the four dams discussed in the Nu River chapter—Maji, Lumadeng, Yabiluo, and Lushui—have a combined budget of 42.4 billion yuan to cover such expenditures. Of this total, 1.8 billion—a little more than 4 percent—is earmarked for resettlement costs (Tullos et al. 2013). Applying the logic of the market, it is not hard to understand why resettlement accounts for such a small share of the budget. To hydropower companies or to central-government authorities tasked with devising and carrying out compensation programs, investing in resettlement is like throwing money away. The expected return on such an investment is minimal or nonexistent. It is no wonder, then, that displaced people around the world tend to be viewed as "people in the way" (Oliver-Smith 2010), whose expedient and discreet removal is a necessary, if regrettable, step in the process of building dams.

A vast chasm separates the powerful decision makers within the hydropower sector and the Yunnan villagers whom their decisions affect. This chasm is geographical—with decision makers occupying glass-and-steel high-rises in Beijing and resettled Yunnan villagers living in newly built "migrant villages"—but it also stems from these two groups' relative positions in the political economy. In effect, such distance allows decision makers to exclude displaced people from the calculus of costs and benefits. Dr. Li's experience with displacement in Yunnan and elsewhere had convinced him that such "distancing" was endemic to the resettlement process because it allowed those with political and economic power to sidestep the moral economy, in effect avoiding a real evaluation of the human costs of their decisions. He reflected, "Displacement is possible only because the hydropower-development companies aren't forced to look at the results of their actions. The hydropower-development companies shouldn't be allowed to keep their distance. They should have to be shut in a room with the villagers that they have displaced. They should have to look those villagers in the eye."

Of course, such distancing is not unique to the dam-building enterprise, nor is it unique to China. Large organizations—governments, corporations, and international financial institutions alike—tend to insulate themselves from the results of their actions by virtue of their elite status. When people are in the way of progress, such people are dealt with through technocratic, economic, or legalistic methods that are often disguised as simply the "workings out of the market" (Oliver-Smith 2010:3). The emphasis on market-driven development in the electricity sector is crucial because market solutions are praised with a kind of utopian idealism in contemporary China. But it would be a mistake to ignore the roles played by various state agencies, too. Contemporary dam building in China represents a blending of the most problematic aspects of both systems: a capitalist market characterized by constant expansion and an externalization of costs onto people who lack political power, coupled with an authoritarian state that can remove many of the obstacles that stand in its way. As the anthropologist Li Zhang has noted in the context of urban-redevelopment projects that displace people while generating profits for development companies and local governments alike, "What further complicates the Chinese situation is the attempt to combine economic liberalization, neoliberal thinking, and socialist authoritarian rule at once. . . . In the national race for wealth and status, a new kind of self-managing, self-directing subject is emerging while the state is gradually shedding its accountability and moral responsibilities for taking care of its people" (Zhang 2010:215–216).

Marketization has undoubtedly proven to be an efficient way to build electrical capacity and to transmit the power to where commercial and residential demand is highest. After all, the market is ideally suited to perform the amoral tasks of matching supply with demand, setting prices, and distributing commodities. In *The Great Transformation*, Karl Polanyi ([1944] 1957) observes that the market sphere has consistently ballooned since the advent of the Industrial Revolution, in the process becoming disembedded from society, less and less accountable to the broader social good, and ultimately unconcerned with questions of morality. But if Chinese decision makers and citizens are also concerned about the collective social good—about the just treatment of people among the most vulnerable in the country, who are asked to bear some significant and lasting costs—then there is clearly a role for the state to play. There is a case to be made that only the state, equipped with political power and ideally accountable to its citizens, can mitigate the "perils inherent in a self-regulating market system" (Polanyi [1944] 1957:76).

MOVING IN THE RIGHT DIRECTION

Given the vast sums of money involved in constructing and operating dams, it is easy to become cynical about government agencies' and hydropower corporations' rhetorical commitment to addressing the needs of displaced people. Indeed, as the detailed case studies of the Nu and Lancang basins illustrate, dam-induced displacement can cause enormous suffering. Such cynicism, in short, is well founded. But it is also true that China's resettlement-compensation policies are gradually improving, due largely to the tireless efforts of Chinese social scientists, journalists, and social activists who have documented the disastrous consequences of displacement and advocated for policies that better address migrants' concerns and better equip them to secure their future livelihoods (see, for example, Guo 2008). The question of precisely how compensation policies affect villagers' lives and livelihoods is ultimately an empirical one.

The structure of law and policy governing population resettlement in China is a complex and evolving mosaic. At its center is the Constitution and the Administrative Law of China; other pieces of the mosaic are formed by laws (*fa lü*) passed by the National People's Congress as well as by administrative regulations (*xingzheng fagui*) passed by state agencies such as the State Council and its various divisions and ministries. There are also local or regional regulations (*difangxing fagui*) issued by local people's congresses or other administrative bodies, which are limited to the scope of their jurisdiction.

From the founding of the PRC in 1949 until the 1990s, an era that witnessed the construction of tens of thousands of dams and the displacement of an estimated 10 million people, there was no nationally concerted effort to include the needs of displaced people in river-basin planning and hydropower development. Progress in such efforts has come slowly and incrementally. Several decades ago, for example, China's resettlement policy was geared only toward addressing "state construction projects," including urban and rural residence infrastructure, industries and mining, transportation, water conservancy, military facilities, and related projects (as indicated in the Land Administration Law of 1998 [Chinese National People's Congress (1986–1998) 2004]). More recently, the policy framework has been expanded to include projects related to other public interests, such as national security and economic development. In practice, this expansion captures a huge range of projects related to energy, transportation, water-conservancy infrastructure, public health, environmental conservation, disaster prevention, historic preservation, and even sports (Chinese State Council 2011).

The two most prominent issues in contemporary resettlement policy, which are also the most politically sensitive and controversial, are demolition for urban-redevelopment programs—often called *chaiqian*, literally "demolishing and relocating"—and resettlement for dam projects.[4] These two issues represent the primary sites of friction arising from rapid economic development, ambiguous property rights, and the transition from a centrally planned economy to a market economy. Table 6.2 provides a description of the most significant laws and policies related to land requisition, resettlement, and compensation in China over the past two decades.

TABLE 6.2 Selected Laws and Policies on Land Requisition, Resettlement, and Compensation in the People's Republic of China

Issuing Body (Year)	Law/Regulation	Description
State Council (1991, revised 1998)	Enforcement Ordinance of the Land Administration Law of the PRC	Offers specific guidelines on how to enforce the Land Administration Law; provides guidelines on public participation in hearings.
State Council (1991, revised 2006)	Regulations on Land-Acquisition Compensation and Resettlement of Migrants for Construction of Large- and Medium-Scale Water-Conservancy and Hydro-power Projects	Outlines administrative rules on resettlement and compensation related to water projects.
NDRC, Ministry of Finance, MWR (1996)	Notice on Establishing Later-Stage Support Funding for Dam Projects	Recognizes the insufficient level of compensation for dam projects commissioned prior to 1995 and establishes a support fund to raise these compensation levels.
Ministry of Land Resources (2001)	Methods for Public Announcement of Land Appropriation	Establishes the guidelines for making public announcement of the intent to appropriate land, the plans for resettlement, and compensation strategies; outlines appropriate timeframes for public notification, planning, and implementation of resettlement and compensation.
Ministry of Land Resources (2004)	Guidance on Improving Land Appropriation and Resettlement	Sets standards for land compensation and procedures for government oversight of land appropriation; outlines procedures for job training and social security services for rural people resettled in urban areas.

TABLE 6.2 (Continued)

Issuing Body (Year)	Law/Regulation	Description
State Council (2006)	Opinions on Improving Later-Stage Support for Large and Medium Dam Project Resettlement	Emphasizes the long-term environmental and social sustainability of dam projects; provides minimal additional financial support to migrants for up to twenty years after resettlement, plus financial support for long-term infrastructure projects and job training.
Ministry of Labor and Social Security (2006)	Notice of Guidance on Job Training and Social Security Services for Farmers in Land Appropriation	Emphasizes the importance of job training and the provision of social security services for farmers affected by land appropriation; requires that these concerns be included in socioeconomic development plans.
National People's Congress (2007)	Property Rights Law of the PRC	Establishes a comprehensive framework for property rights in China; states that housing and real estate held by individuals and collectives can legally be appropriated for public interest.

These laws and policies represent considerable evolution over the past two decades. Some differences between the 1991 and 2006 versions of the Regulations on Land-Acquisition Compensation and Resettlement of Migrants for Construction of Large- and Medium-Scale Water-Conservancy and Hydropower Projects are especially noteworthy (see Chinese State Council [1991] 2006). First, two major forms of compensation—one for land requisition and one given as a resettlement subsidy—have been markedly increased. The current standard stipulates that compensation be calculated by assessing the average annual income of a household over the three years prior to displacement, then multiplying that figure by sixteen. Although the 1991 policy delineated separate standards for land compensation and resettlement subsidies, the 2006 policy simply stipulates a total maximum, presumably to allow for greater local flexibility in how the ordinance is applied. It also includes the stipulation that compensation standards might further increase in the future if resettled households find it impossible to maintain their original living standards.[5]

Second, the ordinance expands the range of properties that may be eligible for compensation, including built structures equivalent to their original scales, quality standards, and purposes. Third, the ordinance requires that compensation for resettlement be included in any project budget. It also outlines a framework for accountability: resettlement subsidies are to be allocated by agreement between county governments and the villages within their jurisdictions, and compensation for land requisition is to be paid directly to affected households.

Finally, and perhaps most crucially, the ordinance specifically deals with transparency and equity in the resettlement decision-making process itself, emphasizing that resettled people have a right to know in advance about development plans affecting them and to participate in the process by attending public hearings or submitting comments. The ordinance also requires that the compensation spending plan be based on a fair survey and assessment of villagers' land and subject to the approval of affected people, though there is little detail suggesting how this should be accomplished. In cases where disputes over compensation arise, the ordinance outlines a process for adjudication and adjustment of compensation levels.

Because these policies are new and continually evolving, it is likely that they are sporadically enforced. Indeed, my own experience during field research found neighbors living next to one another in Xiaowan's resettlement village who were compensated at drastically different levels and had very different feelings about how the resettlement process had unfolded. It may be impossible at this point to understand how well such policies are being implemented on a national level. However, the socioeconomic survey data can give us a glimpse into recent changes in compensation patterns and the consequences for household livelihoods. How likely are these programs to alleviate the worst social and economic

TABLE 6.3 Resettlement Compensation Levels by County on the Lancang River

County	Household Count	Dam Site (status)	Average Total Compensation (yuan)
Fengqing	29	Xiaowan (completed 2010)	11,280a
Yun	38	Manwan (completed 1996) Dachaoshan (completed 2003)	4,957a
Lancang	55	Nuozhadu (under construction)	31,420b
Total Sample	122		18,390

Note: Compensation figures were adjusted for inflation to 2010 yuan, using a consumer price index from the World Trade Organization. In the compensation column, the letters *a* and *b* indicate counties whose average compensation levels differ from one another at the $p < 0.05$ level of significance as determined by a one-way analysis of variance (ANOVA) with post hoc tests (Dunnett's C).

problems caused by displacement? In order to address this question, I performed a separate analysis on all households in the socioeconomic survey data set (Brown and Tilt 2010) that reported any income from displacement compensation (*buchang*) over a twelve-month period ($n = 122$). The results are shown in table 6.3.

The difference in means between the three counties was statistically significant, driven primarily by the fact that average compensation levels in Lancang County were statistically far beyond those in Fengqing and Yun Counties. The quality of these data is likely hampered by the political sensitivity of the topic: as I pointed out in the chapter on the Lancang, nearly 250 households in the data set belonged to the resettled category, but only 122 were willing to answer questions about the monetary value of compensation they had received. It is difficult to know how to treat the missing households, whether they reflect uneven compensation or simply uneven reporting of compensation, because the political sensitivity surrounding resettlement can abruptly end conversation and make it difficult to pose follow-up questions.

Despite this limitation, these data provide one of the only available glimpses into how compensation levels vary across different policy eras. In Yun County, villagers displaced during the 1990s and early 2000s for the Manwan and Dachaoshan projects have walked a very difficult road; their compensation levels are far below those of more recently resettled households (see also Galipeau, Ingman, and Tilt 2013). Some villagers reported receiving only a few hundred yuan in total compensation; many resorted to scavenging in garbage bins for food and clothing, a plight that has been well documented (Guo 2008). Furthermore, during the late 1980s and early 1990s, as the nation was undergoing a massive transition from a centrally planned economy to a market economy, government economic planners gradually allowed prices to fluctuate for a variety of commodities, including important construction materials such as wood, steel, and cement. For resettled villagers, this fluctuation translated into serious difficulty using their compensation funds to build houses of comparable quality to the ones they had lost. The villages of Goujie and Wangjiang were particularly hard hit; in these two villages, poor-quality construction and recurring landslides resulted in seventeen houses being declared too dangerous for habitation. These unlucky households were resettled twice.[6]

But the Xiaowan and Nuozhadu Dam projects provide evidence of improving compensation practices. Resettlement for these projects—in Fengqing and Lancang Counties, respectively—was begun in the late 2000s and is ongoing in some locations. Households being resettled for the Nuozhadu project are receiving about eight times more compensation funding than the households at Manwan received roughly two decades earlier, even adjusting for inflation. Some households in Lancang County reported receiving as much as 100,000 yuan (around U.S.$16,000) in total compensation. This represents financial capital that can be used for new housing, educational expenses for children, business investment, or job training. It is the seed money from which people can rebuild their lives.[7]

In 2009, I had an opportunity to walk through the housing development for resettled people in Fengqing County near the Xiaowan Dam site, which locals called the "migrant village" (yimin cun), where several hundred families now lived in two-story cement houses. The living conditions were reasonably good; many of the families had only

FIGURE 6.1 Newly constructed housing for resettled villagers in Xiaowan Township.

recently relocated and were still decorating their new houses. One family had carefully hung a red calligraphy banner from their front door that invoked the protection of the Buddhist goddess of mercy; it read, "People in the care of Guanyin" (*Ren zai Guanyin yanghe zhong*). Many of the houses were equipped with garages, which, as far as I could see, were filled with farm tools and bags of fertilizer rather than with automobiles. The villagers who lived in the resettlement village now had easy access to the township market center, although many were forced to travel considerable distance, sometimes several kilometers, to work on their land plots.

It is difficult to predict how the lives of resettled families will change over the long term. What is clear, however, is that the different outcomes seen in the data—resettled households in Lancang County receiving monetary sums many times greater than those resettled earlier—represent much more than the machinations of the market economy. These data show a clear and deliberate application of state policy to address the widespread poverty, joblessness, and other social problems associated with dam-inducted displacement. Compensation programs, properly conceived and equitably conducted, can have a dramatic effect on people's lives.

7

A BROADER CONFLUENCE

Conservation Initiatives and China's Global Dam Industry

ONE ANALYTICAL thread running throughout this book relates to the moral economy of water and energy as well as to the relationships between different constituent groups that vie to accomplish disparate management goals in the water sector. Scholars working in the moral economy framework commonly point out that the process of economic globalization has changed the game in fundamental ways. In particular, the current era of neoliberal development—which favors market mechanisms over government control and advocates for economic deregulation and cutting public expenditures on social programs—has reconfigured the roles of states and markets, effectively disembedding the market from society. This trend has captured the attention of social scientists. In a recent issue of *American Anthropologist* devoted to the moral economy concept, Marc Edelman argues: "Whether the state has weakened under globalization or simply assumed new functions, it is often no longer the principal focus of the counter-movement to the market. New supranational governance institutions—such as the WTO, IMF, and World Bank—have become major targets as well" (2005:337).

Chinese government agencies and hydropower corporations, faced with new social, political, and environmental challenges in the domestic sphere, are not immune to these transnational forces. The disembedding of the market from the state, driven by globalization, involves two related stories: one in which global organizations become increasingly important players in conservation and river-basin-management efforts within China; and one in which the Chinese dam-building industry, armed with the best expertise and financing in the world today, looks outward to assume a broader role in development initiatives and resource extraction in lesser-developed countries and regions. In the process, various boundaries—political, economic, cultural, and even biological—become blurred. In the struggle to protect sensitive ecological areas, who can legitimately make a claim on the rich biological resources of northwest Yunnan? And what are the geopolitical, environmental, and social implications of Chinese hydropower corporations, equipped with financial backing and diplomatic clout from the government, participating in dam-development initiatives around the world, from Southeast Asia to Latin America and sub-Saharan Africa?

GLOBAL CONSERVATION EFFORTS IN YUNNAN

The contrast between the Lancang and Nu projects—the former begun in the late 1980s and the latter mired in controversy for more than a decade now—highlights the increasingly important role played by international conservation organizations in contemporary China. The Lancang River dam projects were well under way before many international conservation organizations were operating in China. The Nu River projects have become a rallying point for both domestic and international conservation organizations, who use science and advocacy to protect the ecologically sensitive areas of northwest Yunnan.

The creation of nature reserves is now the key strategy used by the central government in biodiversity-conservation efforts; it is also a lever through which environmental NGOs seek to influence government policy. China's policy framework for protected areas, which borrows explicitly from global conservation models, has come to represent a key point of friction between environmental-protection efforts and development initiatives. Government agencies at various administrative levels currently manage more than 2,000 nature reserves throughout China, with a total land area of more than 100 million hectares, nearly one-tenth of the nation's territory (Xu and Melick 2007; Coggins 2003). Northwest Yunnan is a veritable mosaic of conservation efforts involving government agencies as well as prominent international organizations. The central government established the Gaoligongshan Nature Reserve in 1986, and UNESCO layered the Three Parallel Rivers World Heritage Area on top of it in 2003, setting aside fifteen protected areas in eight clusters totaling nearly 1.7 million hectares (UNEP 2009). However, UNESCO's financial investment in the area is quite minimal, and day-to-day management of the protected areas is overseen by prefectural and county government agencies; these cash-strapped agencies often rely on tourism revenue to support conservation activities (Grumbine 2010).

According to one UNESCO document, "the 1.7 million hectare site features sections of the upper reaches of three of the great rivers of Asia: the Yangtze (Jinsha), Mekong (Lancang) and Salween (Nu River) run roughly parallel, north to south, through steep gorges which, in places, are 3,000 m deep and are bordered by glaciated peaks more than 6,000 m high" (2003:1). After the approval of the thirteen-dam cascade project was first announced for the Nu River, the UN World Heritage Committee issued a warning to the Chinese government to express its "gravest concerns on the impacts that the proposed construction of dams could have on the outstanding universal value of this World Heritage property" (2004:1). In response, the Yunnan provincial government stated that the World Heritage designation comes into effect only at an altitude of 2,000 meters and that the protected areas therefore exclude most of the Nu River Gorge itself (Mertha 2008).

Pudacuo National Park in Shangri-La County, which is one parcel of the Three Parallel Rivers World Heritage Area, is the first park in China to meet the standards of the International Union for the Conservation of Nature, a conglomeration of governments, conservation NGOs, scientists, and industry representatives from nearly 200 countries whose mission is to "conserve the integrity and diversity of nature" through science and policy advocacy (Dowie 2009: xvi). The park area centers on two high alpine lakes—Shudu and Bitahai—both of which are situated at about 3,500 meters above sea level.

On a recent early-summer visit to the national park, which is located only 30 kilometers outside of Shangri-La Old Town, I saw a beautiful dwarf rhododendron species with purple flowers in bloom. The shorelines of both lakes were lined with thick stands of birch and other deciduous trees, the higher elevations covered in larch, spruce, and fir as well as a dense understory of rhododendron, azalea, and wildflowers. The valley bottoms, showing the telltale U-shaped morphology of recently retreated glaciers, were carpeted in closely cropped grassland that was still being grazed by roving herds of yaks and dzos; a local herder outside the park entrance told me that although human settlement is not allowed inside the park, local families are permitted regular pasture access. Chinese and foreign tourists, who can often be seen squinting in the intense sunlight and inhaling supplemental oxygen from canisters, pay a considerable entry fee to spend the day shuttling through the park by bus or strolling along boardwalks that skirt the edges of the lakes and lead to prominent overlook areas. Nearby restaurants have adapted the local cuisine to a variety of palates: stir-fried yak meat for the Chinese tourists, yak burgers for the foreigners.

As I have already noted, the conservationist project is not new to Yunnan; European and American explorers and naturalists have been cataloging, extracting, and even seeking to protect the biological treasures of Yunnan for more than a century. But the extent of global involvement now evident in Yunnan is truly remarkable, as is the extent to which Yunnan's conservation models have borrowed so explicitly from the West (Weller 2006). Table 7.1 shows the categorization schema used by the International Union for the Conservation of Nature to designate different types of protected areas. Central to these global conservation initiatives are semantic and symbolic constructions such as *biodiversity*. The term usually refers to species richness, which, depending on spatial scale, can denote the number of species in a habitat, in a landscape, or in an ecoregion. But in the minds of those who operate conservation organizations and in formal mission statements or policy proposals, alternative definitions of the term *biodiversity* are also put forward. Sometimes taxonomic uniqueness—the number of endemic species in a given area—is a primary rationale for conservation efforts. Other times, the presence of keystone species that perform crucial ecological functions or the presence of high-profile species—so-called charismatic megafauna—become rallying points for international conservation efforts, attracting donations or encouraging volunteerism.[1]

Conservation biologists have warned that we are now living through the "sixth great extinction" in the roughly 4-billion-year history of life on earth, from the rise of single-celled organisms to the ascent of human beings with frontal cortexes more complex than anything in the known universe and endowed with the unique capacity to radically alter the biosphere. Unlike the previous five mass-extinction events—the last of which caused the demise of the dinosaurs approximately 65 million years ago—the current extinction is the unique product of anthropogenic forces such as habitat destruction, species translocation, overpopulation, pollution, and the unsustainable extraction and use of

TABLE 7.1 International Union for the Conservation of Nature Categorization of Protected Areas

Category	Description
Category Ia: Strict Nature Reserve	Restricted from all human disturbance except scientific study, environmental monitoring, and educational activities.
Category Ib: Wilderness Area	Restricted from most human disturbance, including roads and motorized vehicles. Makes provisions for indigenous people practicing traditional livelihoods to access resources.
Category II: National Park	The main objective is to preserve ecosystem functions, while also allowing human visitation, tourism, and educational activities.
Category III: Natural Monument	Designed to protect natural or culturally significant landscape features, with cultural preservation providing an incentive for environmental protection.
Category IV: Habitat- or Species-Management Area	Focuses on conservation of a threatened species or degraded habitat, often involving ecosystem restoration.
Category V: Protected Landscape	Sets aside land areas for conservation but also permits a wide range of human economic activities, such as agriculture, forestry, and ecotourism.
Category VI: Protected Area with Sustainable Use of Natural Resources	Allows a wider range of economic activities but encourages maintenance of the land in accordance with its "natural condition."

Source: Dudley 2008.

natural resources. Such rapid loss of biodiversity—at a rate that likely tops thousands of species per year—represents what the eminent entomologist and popular-science writer E. O. Wilson (2007) has called the "pauperization of earth." Biodiversity loss diminishes the integrity of earth's ecosystems and, in so doing, may ultimately lead to the collapse of our own species.

The establishment of protected areas of varying scales and purposes constitutes one of history's most significant resource-management changes, a transition that is driven largely by the discourse of biodiversity. In this sense, the move to set aside areas for conservation in China represents part of global trend: at present, more than 100,000 protected areas exist worldwide, covering more than 20 million square kilometers, or approximately 12 percent of the world's land surface (West and Brockington 2006). Such a strategy has been called "fortress conservation" (Brockington 2002) because of its exclusion of human activities, generally through the use of law and policy rather than through the construction of actual walls or fences, for the purpose of biodiversity conservation.

Proponents of fortress conservation argue that the ecosystems targeted for conservation have long been threatened by human activity and must therefore be restored to their original, "pristine" conditions. The biodiversity conservation discourse is bolstered by a series of related concepts—diversity hot spots, threatened species, vulnerable ecosystems—ideological constructions that tend to view natural systems as bucolic wilderness rather than as tended landscapes that are the result of generations of human occupation and sometimes intensive management. Henry David Thoreau, the American naturalist and author whose thinking was foundational to the transcendentalist movement in which urban people, stripped of their attachment to nature, sought to reclaim morality and spiritual refuge in wild places, once famously declared, "In wilderness is the preservation of the world."

The fortress-conservation idea came to full fruition in America through the ambitious efforts of iconic figures such as John Muir and their attempts to protect some of the enigmatic landscapes of the American West, including Yosemite Valley and Yellowstone. But the idea quickly grew legs, spreading to Australia, Africa, and Asia, often in step with colonial occupation. Biodiversity is now arguably the chief organizing concept of conservation efforts worldwide. It is simultaneously a scientific imperative to understand the complexity of natural systems through interrelated fields such as botany, ecology, and genetics; a drive to capitalize on biological agents, often as commodities for the pharmaceutical industry; and a conservation target that resonates with donors, volunteers, and resource managers (Harper 2002:33).[2]

As I pointed out in detail in the chapters on the Lancang and Nu River basins, Yunnan's natural areas are freighted with epistemological

TABLE 7.2 Fauna in the Three Parallel Rivers Region

Class	Species	Exemplar/Charismatic Species
Mammalia (Mammals)	173 species, 81 endemic	Yunnan snub-faced monkey (*Rhinopithecus bieti*)
Aves (Birds)	417 species, 22 endemic	Yunnan white-eared pheasant (*Crossoptilon crossoptilon*)
Reptilia (Reptiles)	59 species, 27 endemic	Big-headed turtle (*Platysternon megacephalum*)
Amphibia (Amphibians)	36 species, 35 endemic	Yunnan mustache toad (*Leptobrachium ailaonicum*)
Multiple Classes (Fish)	75 species, 35 endemic	Yunnan catfish (*Pseudexostoma yunnanensis*)

Source: Original documentation to establish the Three Parallel Rivers Protected Areas (UNESCO 2003).

significance; an idealized, conceptual landscape overlays the actual, physical one. In making the case for the Three Parallel Rivers World Heritage Area, for example, UNESCO officials referenced the renowned biological diversity of the region, including more than 6,000 plant species and an astounding array of fauna, some 200 species of which occur nowhere else on earth (see table 7.2). Some, such as the Yunnan snub-faced monkey, a highly social primate that lives in forests up to 3,000 meters in elevation and subsists on leaves, fruit, and lichen, are exceedingly rare and have inspired full conservation campaigns in their own right. In doing so, UNESCO and other organizations have effectively staked a global claim on Yunnan's biodiversity, advancing a moral vision of conservation as a critical project in which global actors and organizations should take the lead.

Especially in the highly developed nations of the world, public realization of the rapid decline of biodiversity has led to an imperative to change course. This realization supplies the premise for a powerful argument in favor of mobilizing people, institutions, and financial resources in the service of biodiversity conservation. In suggesting that biodiversity is a global resource that belongs to all of us—even those of us who will never see it with our own eyes, much less rely on it for our subsistence—as a form of common heritage, international conservation organizations raise crucial questions for debate. To whom does biodiversity "belong"? Who can legitimately engage in biodiversity protection and at what spatial scale? In an important sense, biodiversity is not merely a central concept within the natural sciences, but also a social construct, an organizing domain for policy advocacy, and a call to action for global citizens concerned about environmental degradation. It is a "transnational movement" (West 2006:25; see also Lowe 2006).

As this movement takes shape, places such as Yunnan have come to be understood as fragile, precarious, and in need of protection, often from the human inhabitants whose economic activities and management practices have historically shaped the ecology of the place. This modern conservation project is not so different from the ones undertaken by scientist-explorers such as Francis Kingdon Ward and Joseph Rock a century ago: to observe, to catalog, to understand, to protect. I found it particularly striking, however, that the educational and interpretive materials at Pudacuo National Park, along with the guides' narratives, focused almost exclusively on the park's natural features—its alpine lakes, its assemblage of flora and fauna—and even on some of its notable geological features. Meanwhile, the cultural and historical aspects of the place largely go unremarked on. Tourists expect to see the trappings of Tibetan culture in and around the Shangri-La Old Town—with its lamaseries, its wood-framed houses, and its stupas draped in colorful prayer flags—but the park itself is a natural space cordoned off from cultural influence.[3]

CAUTIOUS ADVOCACY AND COMPROMISE

The Nu River case—where contestation between government agencies, hydropower corporations, international NGOs, and local communities has raged for more than a decade—has illustrated the limits of popular environmental activism in China. This is one of the remotest areas of the nation and one marked by poverty and illiteracy. As our Nu River survey research indicates, only about one-third of Nu River villagers whose lands and homes may be inundated have any real knowledge of the projects or the time lines within which they will be carried out (see also Mertha 2008). This lack of information among people most affected raises several fundamental questions about environmental policy as it relates to conservation: What does public participation in environmental advocacy look like? As Arthur Mol and Neil Carter (2006) have noted, China's environmental governance process is gradually changing to include more participation by civil society groups. These groups include a proliferation of international NGOs as well as domestic NGOs and advocacy groups currently numbering more than 3,000. However, in China it can be difficult to unpack the designation *nongovernmental organization* (*feizhengfu zuzhi*), a somewhat threatening-sounding term that political officials and NGOs themselves generally eschew in favor of the more benign word *social group* (*shehui tuanti*). Indeed, the structures, objectives, and operating strategies of such groups can be quite eclectic. Yang Guobin (2005) categorizes environmental NGOs into seven groups on the basis of their formal registration status with the government: (1) registered NGOs; (2) nonprofit enterprises; (3) unregistered voluntary groups; (4) Internet-based groups; (5) student environmental associations; (6) university-affiliated research centers; and, most paradoxically, (7) government-organized NGOs, or GONGOs.

Environmental NGOs face a range of problems in China, including the requirement to register under the formal sponsorship of a government agency; a limited political and legal framework that requires them to tread carefully or risk being closed down; and, like environmental NGOs everywhere, a critical shortage of funding for their operations (Tang and Zhan 2008). They have also historically been barred by legal statute from gaining standing, the legal right to file lawsuits. In 2012, the Civil Procedure Law was amended to allow "government departments and concerned organizations as designated by law" to engage in public-interest litigation, and this will likely result in a dramatic increase in the volume of environmental lawsuits in the near future (Ngo 2012; see also Wang 2007).

I have already alluded to the local activism that has taken place in the Nu River basin, which has remained sporadic and small in scale, due to a lack of information about how projects are proceeding, a lack of capacity to mount a campaign in the face of economic and cultural marginalization, and the high political risks involved in any opposition strategy. Chinese NGOs, however, have been quite vocal on the topic, though they must generally approach policy advocacy in strategic and circumspect ways or risk serious political consequences. As a case in point, Green Watershed, the first Chinese NGO with a specific focus on water-resource issues, was founded by Dr. Yu Xiaogang in 2002 with a mission of representing the rural people whose livelihoods and cultural practices are intimately tied to southwest China's rivers. Yu, who was educated at the Asian Institute of Technology in Thailand, worked as a researcher at the Yunnan Academy of Social Sciences; he also had a long track record of collaboration with Western NGOs, including The Nature Conservancy and the Ford Foundation. Unlike many of his academic peers in China, he used concepts such as "indigenous identity" and even "comanagement" to advocate for the rights of local people, which likely further alienated him from political officials who continued to toe the official party line that these groups constitute "minority nationalities" rather than "indigenous people" (Hathaway 2013, 2010).

In a bold move, Green Watershed organized trips for villagers near the Nu River dam sites to visit the Manwan Dam on the Lancang River, where they could see for themselves how resettled villagers lived, a picture that was likely quite unfavorable. Yu and his colleagues also arranged for Nu River villagers to voice their concerns at an international hydropower conference in Beijing that was jointly sponsored by the UN and the NDRC. This move raised the ire of both private hydropower-development interests and NDRC officials and ultimately resulted in the temporary revocation of Yu's passport. Green Watershed also came under close scrutiny by central and provincial government authorities.

Domestic activism in the Nu River case was aided by the China National Radio journalist and environmental activist Wang Yongchen, who founded a Beijing-based organization called Global Environmental Volunteers. She organized a petition in opposition to the Nu River dams that was signed by leading figures in science, journalism, the arts, and environmental protection (Mertha 2008). A major part of Wang's work involved alerting the general public to the fact that local people in the Nu River watershed were neither informed nor consulted about the projects (Chen 2006).[4]

Hydropower corporation officials responded swiftly and publicly to this affront by enlisting Dr. Fang Zhouzi, a scholar from the Chinese Academy of Sciences, who gave a speech at Yunnan University with the unsubtle title "A Direct Attack on Fake Environmentalist Dam Opponents," in which he lambasted antidam activists for failing to understand the scientific and technical merits of the case and for engaging in groundless activism (McDonald 2007; Magee 2006). As often happens in cases of dissent in China, Yu's work received a warmer reception abroad than at home: in 2006, he was awarded the Goldman Environmental Prize, a prestigious international award given to grassroots environmental activists, but he remains a controversial figure.[5]

Although it is tempting to view the advocacy movements that surround the Nu River projects as coming from the "grass roots"—particularly since Dr. Yu was awarded the Goldman Prize for grassroots environmental activism—the picture is far more complicated.[6] As Ralph Litzinger has noted,

> To see the opposition as a grassroots movement, the term would need to be expanded to incorporate the activities, mobilization strategies, and media skills of various international and metropolitan-based environmental groups, and the ways in which these groups worked across national, regional, and local administrative boundaries to bring pressure to bear on the Chinese government. Even if we take a more transnational approach to the analysis of the grassroots, we would have to acknowledge that intense debates still rage as to just how much this activism across borders is giving voice to the people living in the Nu River valley.
>
> (2007:292–293)

International organizations have arguably played a much more visible role than domestic NGOs or citizens' groups in the conservation debates taking place over northwest Yunnan. For example, International Rivers, an NGO that began in the 1980s, enlists natural and social scientists from around the world in the effort to preserve free-flowing rivers; the organization added its first full-time staff member in China in 2012. It works by supporting—sometimes monetarily and sometimes logistically—the work of domestic scientists, activists, and NGOs. Other organizations such as the China Rivers Project and Last Descents have focused on publicizing the recreational and touristic value of China's great rivers. The latter organization promotes hands-on environmental education by organizing rafting expeditions that allow people—sometimes those with policy clout—to experience life on the river, hoping that this kind of intimate experience will help people to see rivers not just in economic terms, but for their ecological and cultural value as well.

In my conversations with many government officials and scientists regarding environmental NGOs, I heard time and again that one organization—The Nature Conservancy (TNC, Da Ziran Baohu Xiehui)—was particularly successful in promoting conservation-friendly policy without antagonizing government agencies. Founded in the 1950s in the United States and funded largely through private-member donations, TNC has grown into one of the largest conservation organizations in the world, with a mission of preserving both terrestrial and marine habitats through a pragmatic strategy that engages governments, businesses, and civil society organizations.

TNC began operating in China in 1998 and maintains offices in Beijing, Hong Kong, and Kunming. From the beginning, northwest Yunnan figured prominently into its conservation strategy, and TNC representatives quickly established field offices in Lijiang, Shangri-La, Deqin, and Gongshan. Chinese and foreign scientists in TNC collaborated closely with provincial and central authorities to establish many of the reserve clusters that later became the Three Parallel Rivers World Heritage Area. In tandem with its conservation efforts, TNC has also promoted community-development and poverty-alleviation programs, alternative-energy projects, and educational outreach activities. This sets TNC apart somewhat from other organizations with a similarly global reach, which detractors have collectively dubbed "BINGOS" (big international NGOs), claiming that such organizations' singular mission to protect vast tracts of land often marginalizes the people who rely on natural resources for their livelihoods.

TNC's effectiveness seems to stem largely from a strategy in which pragmatism trumps ideology. I interviewed Mr. Xu, a TNC staff member in Beijing whose work focuses on freshwater ecology. The principal focus of the Freshwater Program is on the Yangtze River basin, but the organization runs projects on many other rivers, too. Mr. Xu, who described his work as a mixture of science, policy, and public outreach, conceded that "TNC believes dams obviously have impacts on natural systems, on the environment." But he went to great lengths to make it clear that the organization also recognized the benefits of dams for energy production and flood control. "Our role," he suggested, "is to help find solutions, to find ways to help minimize the negative impacts."

As he explained the approach his organization takes to policy advocacy, it became clear that two basic characteristics of TNC were fundamental to its success. First, the organization tends to be nonconfrontational in its advocacy campaigns. As Mr. Xu pointed out, "If you put yourself in opposition to development projects such as dams, you lose the opportunity to help improve performance. We will not always say yes to dams,

but we will not always say no, either." Second, in line with the overall culture of the global organization, TNC in China promotes a science-based
approach to conservation that combines technical solutions, policy innovation, and frequent compromise. In the arena of dams, TNC invests a
great deal of scientific effort in understanding "environmental flows"—
sometimes called "e-flows"—or the practice of maintaining a flow regime
at a given dam facility that provides the benefit of hydropower while also
meeting key environmental requirements such as water quality and availability, fish habitat, temperature, and sediment load (Postel and Richter
2003). This goal is often accomplished by operating dams in such a way
as to mimic the river's natural-flow regime. Reflecting on his organization's use of science-based advocacy, Mr. Xu remarked:

> You have to understand, from a scientific perspective, what the poten-
> tial impacts are; then you can find ways to mitigate them. We need to
> really understand the fields of dam engineering and operation, then we
> can make our recommendations on how to manage a dam. For exam-
> ple, we will often send a recommendation to the Yangtze River Water
> Resources Commission about how to operate [one of their dams] for
> obtaining the most benefits for conservation. Last year [2011] was the
> first year that the Three Gorges Dam released water for environmental
> flows, and the data show a positive effect on fish. This year [2012] will be
> the second time that the Three Gorges Dam will release water for envi-
> ronmental flows. I think it will become routine in the future.

Taking this kind of pragmatic approach, TNC acknowledges the
authority of government agencies to manage water but also uses science
to advocate for better outcomes, such as e-flows. TNC scientists make
recommendations about maintaining certain flow rates, releasing water
at times when it can most benefit fish, or maintaining a certain sediment
load. In Yunnan, TNC has also been a key player in the assessment of
ecoregions—land areas that contain a geographically distinct assemblage
of flora, fauna, and ecological processes—with the goal of mapping the
distribution of biodiversity and providing policy makers with a set of conservation priorities. Such efforts help to minimize the costs and maximize the benefits of spending money on conservation and to answer the
crucial question: "If our aim is to conserve the most biodiversity with
limited funding, where should we invest?"[7]

Like most global conservation initiatives, TNC's efforts involve a great deal of financial planning. One innovative tool currently under development is what the organization calls the "hydropower-sustainability fund," which would take money from TNC or other NGOs as payment for releasing more e-flow water and allocate that money for projects such as enhancing floodplain infrastructure or restoring wetlands. The strategies used by TNC are often pragmatic, aimed at mitigating the most serious environmental impacts of dams. In the process, the organization avoids being sidelined in the high-level discussions and debates about dams, a common fate of many NGOs operating in China.

INTERNATIONAL RELATIONS ON TRANSBOUNDARY RIVERS

Dam projects on transboundary rivers such as the Lancang and the Nu are subject to intense political scrutiny at the international level. Downstream riparian countries—mostly notably Thailand—have been fairly vocal in their opposition to dam projects on both the Lancang and the Nu, citing environmental impacts such as water availability and water quality as well as social and economic impacts stemming from changes in water availability for irrigation and fish stocks.[8] In 2010, a Bangkok-based organization called Save the Mekong wrote a letter to the Mekong River Commission (MRC), the intergovernmental agency that works for shared governance of the watershed, pleading for more transparency in its scientific and policy-advocacy activities. The organization requested that the MRC release data on reservoir storage practices at key dam sites, especially Xiaowan, into the public domain. They also asked that the MRC make public any data-sharing agreements that exist between it and the Chinese government (Save the Mekong Coalition 2010).

Such calls have only intensified in recent years as southwestern China and Southeast Asia have faced the most serious and prolonged period of drought in a century. Between 2010 and 2013, news of the drought seized national headlines: farmers suffered billions of dollars in economic damages from lost crops; millions of villagers lost reliable access to drinking water and irrigation; and shipping was halted temporarily on the Lancang because the depleted flow could not accommodate vessels with deep drafts. By coincidence, this period of drought overlapped with much of my fieldwork in Yunnan, and I became accustomed to seeing parched agricultural fields and dry riverbeds throughout the province. So did Chinese citizens around the country: national news on the television and Internet regularly featured stories about the drought and its impact on farmers and residents in the southwest region, illustrating the reports with poignant photographs of agricultural fields dried and fissured for lack of rain.

Speculation abounded that water shortages in the lower Mekong basin were more than simply the result of drought conditions and that they were instead linked to water impoundments behind the Lancang dams. Xiaowan Dam, which has the largest reservoir capacity of any of the dams on the Lancang and which was completed just prior to the onset of the drought, was seen as a primary culprit. Mr. Qin Gang, a spokesman for the Chinese Foreign Ministry, stated publicly that China was committed to maintaining positive relations with other countries in the Greater Mekong Subregion, countries that he referred to as "China's good neighbors" (*China Daily* 2010). But in the absence of specific data on upstream flows and water impoundment, suspicion continued among downstream governments, NGOs, and citizens.

These problems have placed an international spotlight on the governance system of the Mekong basin, which involves China, Myanmar, Laos, Thailand, Cambodia, and Vietnam. Formal, transnational cooperation on the Mekong goes back more than a half-century when, in 1957, the Committee for Coordination of the Investigations of the Lower Mekong Basin—commonly referred to as the "Mekong Committee"—was formed, with Laos, Cambodia, Thailand, and South Vietnam as member states. Cambodia's political instability and civil war kept that country from participating for nearly two decades, but it was readmitted in 1991. The governments of these four nations signed the Agreement on the Cooperation for the Sustainable Development of the Mekong River Basin in 1995, which formally established the MRC with Laos, Cambodia, Thailand, and Vietnam as members (Feng and Magee 2009:109). In 1996, China and Myanmar became "dialog partners" with the commission, but they are not full, participating members.

In 2010, a small delegation from the MRC countries came to the United States under the auspices of the State Department's Lower Mekong Initiative to learn about current best practices in watershed management on transboundary rivers. I sat in on some of the meetings in Oregon, where conversation shifted back and forth from the binational agreement between the United States and Canada governing the Columbia River watershed to the challenges facing the Greater Mekong Subregion. One delegate from Cambodia, a young woman studying for a master's degree, pointed out the importance of collaboration: "We have to make links between local populations, academics, managers, and government agencies. . . . Sometimes, when there are multiple data sources, we don't know whose data to trust." The delegate from Laos, expressing some frustration, concurred, pointing out that "some data is [sic] difficult to share between governments because it's political."

The elephant in the room, which few delegates explicitly acknowledged in these meetings, was the fact that China occupies the headwaters and upper drainage areas of the Mekong but does not participate in many of the MRC's initiatives. Downstream governments and residents perceive issues of water scarcity and water quality downstream—from the Mekong delta in Vietnam to the middle reaches of the river in Cambodia and beyond—as connected to the Chinese dams.

Seasonal fluctuations in the Mekong's hydrograph—its rate of flow over the course of a year, driven by the monsoon rains—regulate a profoundly complex riparian ecosystem. Nowhere are the effects of an altered hydrograph more apparent and more acutely felt than at Tonle Sap, Southeast Asia's largest freshwater lake. Located in central Cambodia, Tonle Sap, whose name means "Great Lake" in Khmer, is a critical migratory bird habitat, an important source of food security in a relatively poor nation that depends heavily on fisheries (Grumbine, Dore, and Xu 2012), and a UNESCO biosphere reserve. Each year during the monsoon floods between May and October, the water levels at the Mekong delta rise high enough to cause the Tonle Sap River to backflow into the lake, quadrupling its surface area. This ebb and flow are the region's "hydrological heartbeat" (Bonheur and Lane 2002) and the mainstay of livelihoods for millions of people.

Since the completion of Manwan Dam, fishermen at Tonle Sap and elsewhere along the Mekong have noticed unusual changes in the river's seasonal hydrograph and a corresponding downturn in fish stocks (Santasombat 2011:40). The potential impacts on food security for the tens of millions of people who live along the lower reaches of the Mekong and who depend on fish as a major protein source and on crop production in floodplain areas adjacent to the river are obvious, but they are also characterized by a great deal of uncertainty. As the Save the Mekong letter made clear, transnational data-sharing arrangements are few, and no one knows much about the volume of water being stored behind the Chinese dams.

Under the operating rules of the MRC, infrastructure projects that affect hydrology are subject to "notification" and "consultation" of the governments of downstream countries; however, there is no effective way for member states to veto projects that they deem unacceptable, even if they are notified and consulted (Santasombat 2011:19).[9] Moreover, China is not the only regional player pursuing its own financial best interest at the expense of its neighbors: in late 2012, the People's Democratic Republic of Laos announced that construction would begin on the Xayaburi Dam in the country's northern region, a project long opposed by other MRC states over fears that it would be the first "domino" to fall. These fears proved justified when in 2013 Laos officials announced the beginning of construction on the Don Sahong Dam, a run-of-the-riven facility near the Cambodian border. River-basin planning documents point to at least eleven potential projects on the Mekong's main stem and more than one hundred on its extensive tributary system; most of the projects entail complex revenue-sharing agreements between multiple riparian countries (Hruby 2013; Ngo 2012).

The Thai anthropologist Santasombat Yos, who has written extensively on the economic and cultural importance of natural resources in the Mekong Region, suggests that Chinese dams in the upper watershed constitute a new form of "transnational enclosure," which he defines as "an increasingly centralized decision-making process which enables the state and commercial interests to gain control of territories that have traditionally been used and cherished by local peoples in the Mekong Basin, transforming these areas into expendable resources for exploitation" (2011:8).

This enclosure process is part of the business of statemaking insofar as it represents the appropriation of resources from peripheral areas and the concentration of political and economic power. It underscores the facts that river-basin sustainability may be threatened by environmental factors such as drought, but that these natural processes are greatly exacerbated by anthropogenic and institutional factors including population growth, infrastructural development projects, and, most crucially, an institutional capacity that is too weak to foster positive change (A. Wolf 2009). During the MRC delegation meetings in the United States, the delegate from Thailand pointed out that in the absence of multilateral cooperation between governments, the role of civil society becomes even more important: "There are different groups—academics, local people, NGOs. We work together. I don't believe the state itself can solve these problems."

In this sense, transboundary governance on the Mekong is a microcosm of an important global issue: approximately 60 percent of the world's freshwater resources flow in 263 international river basins (Wolf, Natharius, and Danielson 1999). Various international laws and policies guide the governance of international rivers, including the Helsinki Rules on the Uses of the Waters of International Rivers (International Law Association 1966), and the Law of Non-Navigational Uses of International Watercourses (UN General Assembly 1997). Both of these statutes call for "equitable and reasonable" use of water by riparian nations, each balancing its own rights against its obligations to neighboring countries. Chinese scholars actively advocate for better upholding of these statutes by sharing data, collaborating with international bodies such as the MRC, and even involving neutral third parties in arbitration when necessary (Feng, He, and Bao 2004).[10]

THE CHINESE DAM INDUSTRY GOES GLOBAL

A corollary to the story of global conservation organizations' engagement in China is the rise of Chinese institutions that possess the will and the capacity to undertake dam-building initiatives around the world. Chinese involvement has been facilitated by some dramatic changes over the past several decades in the global institutional framework for hydropower development. For most of the latter half of the twentieth century, the World Bank held an undisputed position as the largest funder and promoter of dam projects around the globe. But the bank came under intense scrutiny for its failure to anticipate and mitigate some of the worst social and environmental impacts of its projects—a common critique of many international financial institutions. As a result, by the 1990s the World Bank had mostly gotten out of the dam-building business, and global financing for dams from all sources was in precipitous decline (Richter et al. 2010). Public opposition to dams reached its zenith with the Manibeli Declaration, signed in 1994 by more than 2,000 NGO representatives from dozens of countries, which called for a moratorium on World Bank lending to support large dams and for reparations to people affected by its projects over the years.

The first casualty of the Manibeli Declaration to attract widespread media attention was the Arun III Dam Project in Nepal, from which the World Bank withdrew in 1995 under intense pressure from international organizations that highlighted the dam's environmental and social costs.[11] Various civil society organizations from around the world opposed the project, including International Rivers, Greenpeace, and Friends of the Earth. Their major concerns included projected cost overruns, questionable long-term financial feasibility, and, most significantly, adverse impacts on the ecology and culture of the Arun basin. The World Bank's pivot away from dam development left many experts wondering how financing would proceed and whether the bank's withdrawal would be the beginning of the end for dam construction as a global development strategy. As Patrick McCully, president of International Rivers and an outspoken opponent of dams, noted around that time,

> Money is needed, lots of money, and the industry is currently having major trouble getting its hands on it. The World Bank, long the single biggest funder of the international dam industry, is retreating from its critics and has cut the number of dams it is funding to well under half of its peak level. Funding from other multilateral development banks and national development agencies is also declining. . . . Faced with a funding crisis, the industry is desperately looking for justifications for public subsidies. The great hope for the industry is that global warming will come to the rescue—that hydropower will be recognized as a "climate-friendly" technology and receive carbon credits as part of the international emissions trading mechanisms under the Kyoto Protocol.[12]
>
> (2001:xvii)

Enter the Chinese dam-building industry. Over the past two decades, Chinese government agencies, state-owned enterprises, and corporations, taking advantage of this financial vacuum, have come together to make China the world's leading financier, engineering hub, and provisioner of expertise related to hydropower-development projects (Imhof and Lanza 2010). In a recent review of media reports and policy documents, Kristen McDonald, Peter Bosshard, and Nicole Brewer (2009) found ninety-three large dam projects around the world with financial or technical backing from Chinese firms or government agencies or a combination of the two.[13] A selection of these projects is shown in table 7.3.

Chinese involvement takes various forms and works through diverse contractual arrangements, but two dominant models have emerged. Under the "equipment, procurement, and construction" model, a given project is entirely in the hands of a Chinese contractor—from the design of the facility to the provision of financing to construction. The "build, operate, and transfer" model, by contrast, allows a Chinese company to finance and build the project and operate it for a set number of years in order to reclaim its initial capital investment and make a profit. When the established time period—often thirty to fifty years—expires, the company turns the hydropower facility over to the host government for operation and maintenance. From the perspective of a host government, this arrangement is not unlike investing in commodities on the futures market: it represents the potential to reap substantial financial benefits down the road—provided such benefits actually materialize, given the uncertainty of future energy prices— with minimal upfront costs. Between these two extremes exists an array of other kinds of arrangements in which Chinese firms participate in a piecemeal fashion, providing expert advice, equipment, financing options, and other services (International Rivers 2012).

Chinese government agencies and corporations are involved in a huge number of hydropower projects in the Southeast Asian region but also increasingly in Africa and Latin America. On the regional front, Chinese firms have found it particularly fruitful to work in Myanmar, where a military junta has controlled the country for four decades, leaving Burmese citizens with few mechanisms to participate in the decision-making process or to block undesirable projects.[14] On the Salween, as the downstream portion of the Nu River is called, a series of five dams is already under way in Myanmar with investment from Thai, Burmese, and Chinese companies, including Sinohydro Corporation, China's largest state-owned enterprise in the energy and construction sectors, which has built a veritable spider web of subsidiary companies (International Rivers Network 2009).

Similarly, the China Power Investment Corporation (CPIC), a state-owned enterprise under the direction of the State Council, which also holds a controlling interest in many publicly traded subsidiary companies, is working to develop a series of seven dams on the Irrawaddy River in Myanmar—the next major watershed to the west of the Nu basin— including the U.S.$3.6 billion Myitsone Dam. CPIC has proposed to use a build-operate-transfer model under which it provides initial financing and construction, in cooperation with several Yunnan-based companies

TABLE 7.3 Selected Dam Projects Around the World with Significant Financial or Technical Involvement by Chinese Institutions

Dam Project	River	Project Details
Hatgyi Dam (Myanmar)	Salween	1,200 megawatts installed capacity, U.S.$1 billion investment (Sinohydro, China Southern Power Grid, CPIC, Yunnan Machinery and Equipment Company)
Sambor Hydropower Project (Cambodia)	Mekong	7,110 megawatts installed capacity (China Southern Power Grid, Guangxi Power Industry Surveying and Design Institute)
Gomal Zam Dam (Pakistan)	Gomal	17 megawatts installed capacity, U.S.$190 million investment (Sinohydro, with additional financing from U.S. Agency for International Development)
Merowe Dam (Sudan)	Nile	1,250 megawatts installed capacity (China Exim Bank, China International Water and Electric, Sinohydro)
Lower Kafue Gorge Power Station (Zambia)	Kafue	750 megawatts installed capacity, US.$1 billion investment (Sinohydro)
Chalillo Dam (Belize)	Macal	7 megawatts installed capacity (Sinohydro, Yangtze River Commission)

Source: Selected and adapted from McDonald, Bosshard, and Brewer 2009.

and foreign-owned companies. Under the most current arrangement, CPIC would operate the dam for a period of fifty years—while selling most of the power to China but also providing some free electricity and revenue to the Burmese—and then transfer ownership rights of the facility to the Myanmar government (Meng 2012).

But recent developments in the Myitsone Dam project have highlighted the financial and political risks at stake under such arrangements. Media reports exposed the forced removal of thousands of people from the dam area, mostly villagers from the Keqin minority, at the hands of the Burmese military. The president of Myanmar announced in

2011 that the Myitsone Dam would be suspended during the tenure of his administration, citing environmental impacts and concerns for the social protection of minorities; this announcement placed China in the uncomfortable position of being outflanked by Myanmar on issues of transparency and popular democracy.[15] For their part, Chinese officials released a report written by several scholars at the Chinese Academy of Social Sciences that accused Western NGOs of stirring up trouble in the Mekong region, arguing that such groups have "severely damaged China's reputation" by exaggerating claims about the environmental damage caused by Chinese firms (J. Li 2013).

Looking farther afield, China's current involvement in dam development in Africa has enjoyed exponential growth; dozens of projects are under way on a score of major rivers throughout the continent. Sinohydro officials estimate that overseas business now accounts for more than one-quarter of the company's revenues and that, of the overseas portion of its portfolio, business activities in Africa are the most lucrative (Sinohydro 2013). One particularly instructive case is the Lower Kafue Gorge Hydroelectric Project in Zambia, a contract worth nearly U.S.$2 billion that was awarded to Sinohydro through a noncompetitive bidding process in 2010. In an editorial commentary published shortly after the announcement that the project would move forward, Michael Tarney, managing director for corporate development at the Copperbelt Energy Corporation in Zambia, expressed optimism mixed with caution. He acknowledged China's expertise in the field today, which made Sinohydro the natural choice as a key partner. But he also urged concessions such as the employment of Zambian construction workers, a financing package that would make electrical power affordable for ordinary consumers, and technology-transfer programs to boost Zambia's domestic technical capacity. Tarney remarked, "Success and sustainability of a robust and stable key infrastructure like energy generation solely depend on technology transfer. China is where it is today not just because of its role as a manufacturing hub for the world but also due to huge investment in technology transfers" (qtd. in *Post* 2010).

Tarney's comments underscore the importance of technology transfers—the process through which technology, skills, and even specific manufacturing techniques are transferred from one country to another, usually at the behest of the government as a precondition for entering into a business partnership. It is true that technology transfer has played a major role in China's development trajectory, particularly during the Reform and Opening period. Although Chinese companies continue to exercise their primary competitive advantage in the global marketplace—cheap labor—they also rely on the government to help establish technology-transfer agreements with global corporations. As a result, in the space of only a few decades, Chinese engineers have gained expertise in automobile design from American and German companies, in high-speed railway and train construction from Japanese firms, and in wind-power production from European firms (Lewis 2013). These gains have undoubtedly helped China to move up the production chain from labor-intensive manufacturing to more value-added domains such as research and development.

The Chinese model of building dams in African countries is predicated on a set of complementary interests between the two sides. China is in constant need of resources—in particular oil, minerals, and timber—to continue its domestic economic expansion, and Chinese firms can enter into quid pro quo arrangements with governments in order to accomplish this. For example, the China National Petroleum Corporation has been investing in Sudan since the mid-1990s even amid armed conflict that has driven away many Western investors worried about the ethical and public-relations problems with such investment choices. China National Petroleum has engaged in oil-extraction activities in exchange for direct investment in infrastructure projects in Sudan, including a refinery and several hydroelectric dams, the largest of which is the Merowe Dam on the Nile. More than a dozen Chinese nationals, many working in state-owned enterprises such as Sinohydro and China National Petroleum, have been abducted or killed in recent years in conflict zones from Sudan to Afghanistan.

There are several ways to view this trend in Chinese–African engagement. One is by considering the economic impact of increased investment in large-scale infrastructure in some of the world's poorest nations and regions. Africa is the continent with the lowest percentage of its theoretical hydropower currently developed, and power shortages arguably exacerbate poverty and prohibit economic growth (Boyd 2012). From this point of view, developing Africa's hydropower capacity may be seen as a step toward a more reliable electricity supply, which is something on which all other development goals—poverty alleviation, disease reduction, and the expansion of entrepreneurial activity—arguably depend. Of course, Chinese corporations also eye the African continent's one billion people as a future consumer market for their manufactured goods.

The legitimate fear on the part of environmental and human rights activists, however, is that Chinese developers have failed and will likely continue to fail to apply even basic standards for environmental and social review. The developers' quid pro quo approach to development, coupled with a policy of noninterference in the domestic affairs of the countries in which they operate, may make even the most ardent critics of the World Bank long for the days of structural-adjustment programs. Such programs have been a standard tool of multilateral agencies, including the World Bank and the International Monetary Fund, for decades, tying aid and loans to a host government's willingness to carry out political and economic reforms. Structural-adjustment programs, buttressed by the faith in liberal economics that permeates the entire development industry, are geared toward creating greater efficiency and transparency in markets and governments by forcing greater fiscal austerity, privatization of land and industry, trade liberalization, and financial deregulation (Rapley 2007:87–133).

By contrast, the new arrangements between Chinese dam developers and African governments may entail a lack of transparency even greater than was the case under international financial institutions. Because of Chinese hydropower companies' general orientation toward corporate profits, most dams are designed to meet the demands of large industries such as iron and steel or cement rather than building a solid electricity-distribution network and supplying poor people with electricity. Mozambique's Mphanda Nkuwa Dam, for example, which was approved by the national government in 2011 and is projected to cost $2 billion, was designed principally to supply power to the aluminum industry and to export surplus electricity to neighboring South Africa, leaving many of Mozambique's poorest citizens with no reliable electricity supply (Pottinger 2012). This type of outcome may not be "development-induced poverty" in the way of structural adjustments, but it is nevertheless "development-perpetuated poverty" produced by a no-strings-attached approach that raises questions about long-term social stability, let alone democracy and justice.

In contrast to their multilateral counterparts, Chinese government agencies and corporations have shown little interest in building state capacity, enhancing civil society, or implementing structural-adjustment programs that seek to reorient the relationship between the state and the market in the countries where they work. Rather, they tend to focus on quid pro quo development projects: a dam here in exchange for mineral-resource concessions there. Africa—which offers both abundant resources and historically weak governmental institutions—seems particularly vulnerable to this sort of development, typified by fortified enclaves of resource extraction that generate capital for global elites while marginalizing local communities (Ferguson 2006).

One potentially positive interpretation of Chinese involvement in dam-building activities in Africa is that such arrangements sidestep structural-adjustment programs, which have wrought incredibly painful changes in the lives of people around the world who have been affected by them. The development industry is replete with examples of the tragic consequences of structural adjustment: people who are forced to cope with the devaluation of their national currency and, by extension, the loss of their savings; workers being downsized in the name of efficiency; and vulnerable populations failing to receive the services they need because of slashed social programs. Given the inhumanity of many of the policies of international financial institutions such as the World Bank, Chinese firms' quid pro quo approach may come to look comparatively humane.

The debate about China's role in the financing and construction of dams around the world is at its core a reframing of the debate between neoclassical and state-led development. Aihwa Ong suggests that "neoliberalism can also be conceptualized as a new relationship between government and knowledge through which governing activities are recast as nonpolitical and nonideological problems that need technical solutions" (2006:3). As concepts such as neoliberalism gain academic traction, there is a tendency to apply this logic to China's recent economic trajectory. However, while the main story lines in this book certainly provide evidence for a nonideological, even technocratic, approach to development, I am skeptical that the development policies enacted by the CCP can reasonably be seen as "neoliberal," which requires a combination of liberal market policies, an eschewing of central economic planning, *and* at least a rhetorical emphasis on individual freedom and liberty (Harvey 2005).

In practice, neoliberal policies are often quite effective at claiming more and more economic and political power for elites. The more one learns about the institutional structure of hydropower in China—with government agencies, state-owned enterprises, joint ventures, and shareholder corporations such as the Five Energy Giants all playing their various roles—the more the concept of "social embeddedness" seems to come into play. After all, Chinese companies are arguably more embedded than ever: they were born and grew to prominence as state-owned enterprises; they rely on the government to provide development permits for domestic projects, often by skirting around the laws that mandate social and environmental reviews; and they increasingly depend on the Chinese government to smooth their entry into foreign markets through bilateral diplomatic relationships and the provision of favorable financing terms. But they are utterly disembedded in other ways, showing little accountability to local people affected by domestic projects and owing little or nothing to the citizens of the foreign countries where they work, beyond whatever resource-trading agreement has been brokered. If we wish to apply a term to this new arrangement, perhaps an appropriate one would be *neosocialism*, which Frank Pieke uses to refer to a "combination of centralization, strengthening, and selective retreat of the state" (2009:10). In short, the government pulls back in some areas—particularly in exercising regulatory oversight or in providing social services—while continuing to actively facilitate marketization and privatization.

There is nothing novel or unique about this model of public–private cooperation, of course. Critics of international financial institutions and of bilateral aid agencies have long observed a common pattern by which a given development agency secures a contract with a foreign government, favors its own companies or NGOs in the provision of subcontracts, and thereby ensures that many of the long-term benefits of development projects remain insular. In the near term, the key question will be how committed Chinese firms are to abiding by internationally established practices for environmental and social safeguards. Sinohydro, the largest Chinese overseas dam-development corporation, recently announced that it would adopt World Bank displacement and resettlement safeguards as its minimum standards (International Rivers 2012:21). These guidelines require consultation with affected parties and full disclosure of development plans as well as the completion of an EIA prior to the beginning of construction. They also set basic standards for enacting resettlement plans and compensating displaced people, managing the long-term effects on the cultural heritage of vulnerable groups, and establishing clear mechanisms to hear and address grievances. Although there is reason for skepticism about how well these policies will be implemented over the long term, International Rivers has optimistically called Sinohydro's announcement "the first time a Chinese hydropower construction company has articulated its policy commitments at this level of detail" (2012:22).

CONCLUSION

The Moral Economy Revisited

THE ABILITY to regulate the flow of rivers—for irrigation, for flood control, and, more recently, for hydropower—has long been a fundamental part of the story of human development. In this regard, the recent boom of hydropower expansion in China, which has seen the construction of approximately half of the world's 50,000 large dams, can be viewed as a change in scale rather than a change in course. China has a history of large, state-sponsored river-engineering projects that stretches more than 2,000 years into the past; such projects have been fundamental to political expansion and human well-being there ever since. At the same time, the large dams under way on the Lancang and Nu Rivers, along with scores of others on the main stems and tributaries of most of China's major river systems, remind us of the complex questions at stake: water is simultaneously a resource that is central to people's livelihoods, a kinetic force capable of producing renewable energy, and a medium through which social and political relations are negotiated—sometimes in contentious ways.

Chinese leaders are cognizant of the unsustainability of continued reliance on fossil fuels, especially coal, for the vast majority of the country's energy needs. Although such dependence will by necessity continue for the foreseeable future, the nation is making strides toward alternatives, which policy makers refer to as "clean energy" (qingjie nengyuan) or "green energy" (lüse nengyuan), with the long-term goal of promoting a "low-carbon economy" (ditan jingji). The national capacity for research and development as well as construction of wind-, solar-, and hydropower facilities is outstripping that of many advanced industrial economies, including the United States. And these energy programs are a fundamental part of the rise of a nation that is far more prosperous than it was a generation ago. Making general predictions about China's development path can be pure folly, but I will hazard one. Twenty or thirty years from now, air quality in Chinese cities will be markedly better than the "unhealthy" or "hazardous" ratings common today. And these improvements will have been driven by the demands of an increasingly affluent middle class weary of dealing with the environmental and health consequences of fossil fuels as well as by large-scale public and private investment in renewable energy projects of various kinds.

On the Lancang River, where the total hydropower capacity is estimated to be 31,980 megawatts, nearly one-third of that total has been developed so far: the Manwan, Dachaoshan, Xiaowan, and Jinghong Dams are currently operational; Nuozhadu Dam is nearing completion; and a handful of other facilities on both the Upper and Lower Cascades are in various stages of planning or construction. With the commencement of construction on the Xayaburi Dam in Laos in 2012, environmental activists fear the beginning of a "domino effect" that will see scores of projects on the Mekong River's main stem and tributaries in the years to come (Ngo 2012). Riparian nations in the Mekong basin will continue to face major governance challenges as they seek to balance energy development, human well-being, and ecological conservation.

Meanwhile, one river basin to the West, the Nu River development plans have spurred a decade-long battle, which now appears to be winding down in the wake of the 2013 announcement by the State Council that at least five of the dams—including Songta in the Tibet Autonomous Region—will proceed under the Twelfth Five-Year Plan (2011–2015). With the installment of a new government, Premier Wen Jiabao—whom many perceived to be a populist leader with a skeptical view of rapid hydropower expansion—stepped down, thus removing the last major political obstacle to the Nu River projects. Organizations such as UNESCO and International Rivers continue to "remind China of its obligation to protect the Three Parallel Rivers Area under the World Heritage Convention" (International Rivers 2013a), but these appeals seem unlikely to meet with much success. The MEP has released guidelines that forbid the construction of dams within protected areas (Brody 2012), but the boundaries of the Three Parallel Rivers Protected Area in Yunnan all come into effect at areas of higher elevation, away from the locations of extant and planned dams, a political move that in hindsight seems designed to expedite hydropower expansion (Grumbine 2010).

THE MORAL ECONOMY OF WATER AND POWER

As the anthropologist Arthur Kleinman and his colleagues observe in a recent edited volume on the moral dimensions of life in contemporary China, one contribution of anthropological research is to provide critical information about "local worlds" of "moral experience" within which decisions are made and negotiated (2011:3). I have suggested that such an examination is equally necessary for the villagers whose lives are upended by these megaprojects as well as for the scientists, advocates, bureaucrats, and policy makers who are charting China's future course of energy development, sometimes in the midst of great conflict.

In taking a moral economy approach to the study of water and power generation in contemporary China, I have for the most part resisted the temptation to advance a particular agenda about what outcomes are "right." Instead, I have tried to elucidate the goals and strategies of key constituent groups as they relate to balancing conservation and development objectives on Yunnan's rivers and to show how these strategies are grounded in moral, cultural, and historical precedents. Key government agencies, including the NDRC and the MWR, view hydropower development as a means to achieve energy security amid the continually rising demand for electrical power. This goal is significant and laudable, with the potential of generating enough electricity to offset and perhaps even render obsolete hundreds of coal-fired power plants that are at the heart of the nation's air-pollution woes. It is also part of a national strategy to dramatically cut greenhouse gas emissions and to reduce the carbon intensity of major economic activities, especially manufacturing. I have argued that the current boom of hydropower development is part and parcel of the statemaking process; it is a means through which powerful political and economic interests can advance further resource claims in the geographical and cultural periphery of the southwest region. Paradoxically, statemaking involves powerful nonstate actors: hydropower corporations and a dizzying array of subsidiaries that seek to develop the region's hydropower potential and distribute electricity eastward on the grid. International investors view the situation as an attractive financial opportunity because putting money into energy conglomerates is essentially betting on the continued expansion of the Chinese economy.

The conservation agenda in Yunnan—driven by well-known multilateral agencies such as UNESCO, international NGOs such as TNC, and an array of domestic organizations, scientists, journalists, and other advocates—draws heavily on scientific studies that show Yunnan to be a key repository of biological diversity. These organizations stake a claim on Yunnan's unique assemblage of flora and fauna as a form of common global heritage. At the same time, various domestic NGOs work to achieve conservation goals in more circumspect ways. The issue of preserving biological diversity and cultural heritage is not without complications because it is unclear at what geographical scale claims to the region's biodiversity can legitimately be made.

The picture gets considerably more complicated when local villagers' needs are taken into account. On the Lancang, where tens of thousands of villagers have been displaced so far and where many more will undoubtedly be displaced in the years to come, the story is one of adaptation in the face of difficulty and uncertainty. Villagers must cope with dramatic changes in access to agricultural land, which affects their subsistence and their ability to produce commodity crops for the market. A primary means of adaptation is to send one or more household members to cities and towns in search of wage-labor opportunities; resettled households, equipped with some seed money from government compensation programs, appear to be taking advantage of the opportunities presented to them by initiating entrepreneurial activities. But resettled villagers also face an altered social landscape in which their networks of interdependence and cooperation are disrupted. Some resettled households have borrowed significant amounts of money from families and neighbors in recent years, likely as a means to invest in income-generating opportunities, a trend that may have serious negative consequences down the road. The comparison of resettlement at multiple dam sites on the Lancang—first at Manwan and Dachaoshan, then at Xiaowan and Nuozhadu—allows us to examine for the first time how changing government policies regarding resettlement compensation have affected the lives of villagers. Such policy changes represent, I argue, one of the bright spots in this study: recently resettled households are receiving many times the compensation sums that their predecessors did, and these sums become the seed money that allows them to tenaciously try to rebuild their lives.

In the Nu River basin, villagers remain among the poorest citizens in the country. Although many villagers lack specific knowledge about the hydropower projects currently being planned and constructed, most actually support the exploitation of hydropower resources for national economic development. Moreover, they view the projects as a potential job-creation strategy, though evidence from similar projects around the globe suggests that long-term economic gains from such projects rarely materialize at the local level. Many households have transitioned from a swidden farming system only in the past generation or two and still rely on a very small income from agriculture, livestock sales, and nontimber forest products such as mushrooms and herbs. Dam development is occurring simultaneously with massive road-construction campaigns that will soon link the Nu River Gorge with more heavily traveled areas of Yunnan, a transition that is already pushing many local residents out of agriculture and into the wage-labor economy. I have suggested that three areas of vulnerability—economic, political, and cultural—will be important to consider as the Nu River projects move forward. The world will be watching to see whether compensation policies are well implemented and whether some of the worst social ills typically seen with development-induced displacement campaigns—deepened economic marginalization, joblessness, and social unrest—can be avoided or mitigated. Can hydropower development be undertaken in a way that transcends the technocratic approach by meaningfully addressing the moral concerns—about disturbing sensitive ecosystems and displacing vulnerable people—that are at the heart of the controversy?

BEYOND CRITIQUE

When we acknowledge the moral dimensions of hydropower development and the complex trade-offs that it entails, we can more clearly see the things that we value and the things that we sometimes willfully ignore. In my interview with Dr. Liu, the environmental scientist who directed a prominent research center, she hinted at a changing value orientation currently under way among scientists and policy makers in China: "The majority of decision makers have an instrumental orientation [toward the environment]. They think, 'If the environment is destroyed, we [people] will be affected.' They consider the environment to be part of quality of life. Right now, I think that we are developing a second kind of value orientation: the social/cultural value of environment and natural resources is becoming important. This idea acknowledges that people's social and cultural lives are inseparable from the environment."

Indeed, my interactions with many experts and practitioners—engineers, environmental scientists, agency officials, and even hydropower corporation representatives—convinced me that the social and ecological costs of dams are already a topic of major concern in China. The question is how to find pragmatic solutions. At a recent international conference on hydropower in Washington, D.C., a senior Chinese scientist from a government agency expressed frustration at what he perceived to be constant criticism of Chinese policy coming from the West on topics as disparate as human rights and environmental policy. "Everyone tells China, 'You're wrong,'" he reflected. "We'll burn fossil fuels for economic development. No, that's bad. Okay, so we will build dams. No, that's bad. Then we'll subsidize renewable energy. No, that's bad." In frustration, he raised a sarcastic, if rhetorical, question: "What are we going to do—turn out the lights?"

He reminded the conference participants of the fact that energy issues have become a major focus of international geopolitics as Chinese manufacturers and exporters of solar panels, for example, have enjoyed generous government subsidies that allow them to "dump" their products on foreign markets at prices that undercut U.S. manufacturers. In response, the U.S. Department of Commerce has placed duties on solar panels imported from China and has filed formal disputes with the World Trade Organization. He suggested that our goal should be "to provide advice rather than criticism."

One productive way forward, in my view, is to steer the conversation toward a basic question: If we are truly concerned about the ecological and social effects of dams, such as those on the Lancang and Nu Rivers, what principles should guide us as we decide whether and how these projects move forward? I would like to suggest several. First, decision makers need to apply better tools that can provide a clear-eyed assessment of the real costs and benefits of a given project. There is no getting around the fact that this will require managing river systems for multiple objectives: energy production and economic development, riparian ecosystem health, biodiversity conservation, and local residents' social and economic well-being.

I have suggested that decision makers often view the true costs of building dams—in economic terms, but also in the toll on social well-being and cultural heritage—as incidental externalities, the *faux frais* of such projects. But the people whose lives are upended by displacement and resettlement must be put at the center of the discussion. Decision makers would do well to pull themselves away from the balance sheets and cost–benefit analyses on their computer screens and spend some time strolling through a resettled village. Villagers, NGO representatives, and even savvy government officials acknowledged to me that the human costs of displacement are becoming increasingly untenable.

One reason for the persistence of these externalities is the political polarization that now characterizes the debate over hydropower. At the same time that government agencies and hydropower corporations promote the dam projects in Yunnan by referring to them, quite disingenuously, as "poverty-alleviation weapons," international conservation organizations bemoan the projects by making references to ecological and cultural "destruction" in China's so-called Grand Canyons. The various parties with an interest in Yunnan's rivers have staked out oppositional positions, marking their ground with fairly extreme rhetoric. In truth, focusing on a single management objective is relatively easy to do. If your focus is on alternative-energy development, you advocate for dam construction. If your focus is on community equity, cultural preservation, or ecological conservation, you oppose the dams. Moreover, each group, anticipating a process of conflict and negotiation, feels compelled to stake out the most extreme position possible in order to avoid losing too much ground. And the loudest voices, usually amplified by money, are the ones that get heard.

But energy development, like most complex problems, is ultimately not reducible to a set of binary or even discrete choices. My participation in cross-disciplinary research on dam development has convinced me that the truly difficult task, the one that must be undertaken, is to conceive of outcomes that maximize more than one dimension—plans for action that, for example, help to reduce dependence on fossil fuels *and* prevent ecological harm and economic and cultural marginalization. Various models are now available—including the Integrative Dam Assessment Model—to aid multicriteria decision making by helping leaders to make a real accounting of the costs and benefits of a given project (see Tullos et al. 2010). Another tool now being put to use by government agencies and energy corporations is the Hydropower Sustainability Assessment Protocol, developed by the International Hydropower Association, a non-profit organization with strong backing from the hydropower industry. Critics have argued that the protocol is an attempt to circumvent the most robust calls by the WCD for social and environmental responsibility (Imhof and Lanza 2010) and that the dam industry, under a deluge of complaints from civil society organizations, has simply devised a "green-washing" cover that will allow them to continue business as usual. That may be. But such decision-support tools ultimately allow policy makers to see more clearly how their choices affect ecosystems and communities. They also serve as a transparent record of decision making, showing where trade-offs are made and rendering explicit decision makers' normative judgments that would otherwise go unexamined.

A second recommendation to move the conversation beyond critique is to advocate consistently what the UN General Assembly's Declaration on the Right to Development refers to as "meaningful participation in development and in the fair distribution of the benefits resulting therefrom" (UN General Assembly 1986) among key constituent groups. Evidence suggests that both state agencies and foreign NGOs in China are adopting participatory approaches with greater frequency, driven in part by central policy that mandates public hearings as part of the guidelines for complying with the EIA Law (Tilt 2011). The World Bank and other multilateral agencies that have long funded hydropower development have established clear guidelines for assessing the social impacts of dam construction in a way that acknowledges the rights of local people and seeks to mitigate the risks they face.

Such a framework is admittedly extremely difficult to implement in China, where governmental policies tend to be top down. As I have outlined, recent revisions to national policies on compensation to displaced populations represent a major step forward from China's historically inadequate compensation structure. However, as is often the case, the trouble lies in the implementation of regulations. The limited resettlement campaigns that have been carried out thus far in the Nu River Gorge show a basic failure to implement key policies such as job training.

Public participation presents a range of challenges. Resettled people often belong to socioeconomic or ethnic groups that have been historically marginalized and vulnerable. Their assets may be nonmaterial or otherwise unprotected by formal land rights. And the system of legal arbitration to which they might turn for protection is weak and difficult to access. Which brings us to a crucial point: hydropower-development projects are afflicted by an ailment that appears almost universal in reform-era China—namely, the lack of a clear and consistent accountability mechanism. Many of the experts I interviewed about their scientific activities to assess dam impacts and their forays into policy advocacy expressed dismay at the lack of accountability. Their carefully collected data, their charts and graphs, and their exhaustively researched and footnoted reports all seemed to disappear into the bureaucratic morass when decisions were made behind closed doors. For villagers whose lives and livelihoods hang in the balance of these hydropower projects, this lack of accountability can be alienating and terrifying. Indeed, many villagers expressed a deep sense of ambivalence: a general confusion about the likelihood that they will be displaced and about the compensation they might expect—a sense of fatalism and cynicism that suggests "nothing can be done [*meiyou banfa*]" tempered by the tenacious hope that the government is watching out for them.

I have discussed a variety of measures commonly pursued by rural people in the face of displacement, especially the practice of *shangfang*, or petitioning. The overall effectiveness of tactics such as *shangfang* is difficult to assess. I have outlined various cases over the past several decades in which disgruntled villagers, seeking redress, "look for an upright official [*zhao qingtian*]" who will hear their concerns and advocate on their behalf. But the precariousness of this tactic was made clear when Premier Wen Jiabao, whom many considered to be such an official, was replaced in 2013, and the Nu River projects were immediately put back on the State Council agenda.

All of this suggests an acute need for political and legal levers to hold government agencies and hydropower corporations accountable for their actions. Two important frameworks—EIA and SIA—have evolved in recent years and may hold great promise, although significant problems in their implementation remain for the time being. Billion-dollar projects sometimes go from design to construction before any review is conducted; environmental scientists, through complacence, corruption, or malfeasance, produce reports of dubious quality; and a convoluted, multiagency approach to IRBM can result in the "nine-dragon" problem in which the lines of responsibility are obscured.

In short, the crucial shortfall that allows social and environmental costs to be externalized or written off is accountability.[1] Procedural statutes such as EIA and SIA admittedly constitute thin ground on which to build such accountability. But to argue that they don't work at all is to ignore the evidence: the Nu River projects were stalled for a decade by prominent officials who invoked the EIA Law in their consistent demands for a cautious approach; and the improved compensation standards on the Lancang River were supported by SIA reports from other dam projects, including the Three Gorges Dam, that exposed social problems and advocated for change.

If progress toward a more just accounting of social costs is to be made, it will require consideration of the most challenging issue in China's economic transformation: rural land tenure. Villagers in Yunnan, like villagers elsewhere in China, enjoy usufruct rights over their land via the Household Responsibility System, but they hold no formal land title, which still belongs to the rural cooperatives, the vestiges of communal farming during the socialist period. Many of the failings of current compensation policies derive ultimately from this sort of institutional indeterminacy: the government recognizes use rights and determines compensation levels based on lost agricultural income, while refusing to recognize ownership or salability of land and failing to acknowledge the importance of informal resource-harvesting rights. This narrow definition of land rights, which is codified in the PRC's Constitution, all but ensures that inadequate compensation will continue. I have suggested that the Five Energy Giants and other energy-development interests benefit from the status quo because government policies ensure that land ownership and salability, which represent the most valuable portion of the land-rights bundle, are beyond the reach of rural villagers and therefore do not merit compensation.

It is worthwhile to consider the following question as a simple mental exercise: How much would it cost to compensate people adequately for displacement from their land and dispossession of their homes? To take the Nu River projects as one brief example, I have noted that the total cost of just four of the dams—Maji, Lumadeng, Yabiluo, and Lushui—runs to 42.4 billion yuan, including 1.8 billion yuan for resettlement costs, a figure that includes compensation given to households. Meanwhile, current estimates suggest that annual revenue from the sale of hydropower may amount to 16 billion yuan. This means that monetary consideration of the human beings who incur the greatest damages from the projects accounts for just 4 percent of the upfront capital expenditures. Moreover, in the decades to come, villagers will see only a miniscule portion—far less than one percent—of the long-term financial benefits derived from the sale of hydropower. This discrepancy seems like an obvious problem that can be addressed with minimal effort. The marginal cost of doubling or even tripling compensation rates, viewed in the light of total capital outlays for the construction of a series of multi-billion-dollar dams, is quite small. To villagers, however, this sum would provide crucial support. Evidence suggests that households who receive higher compensation are more likely to engage in entrepreneurial activities, start businesses, or otherwise engage in the process of rebuilding their lives.

KNOWLEDGE, APPLICATION, AND ENGAGEMENT

In a recent speech to the Association of Southeast Asian Nations, then–U.S. secretary of state Hillary Clinton specifically addressed dam construction in the Mekong Region, invoking the American experience as a cautionary tale: "I'll be very honest with you. We made a lot of mistakes. . . . We've learned some hard lessons about what happens when you make certain infrastructure decisions and I think that we can all contribute to helping the nations of the Mekong region avoid the mistakes that we and others made" (qtd. in *The Economist* 2012).

Throughout the process of learning more about the history of large dams as a development strategy, I have been fascinated to see how many of the historical precedents point back to the United States and its efforts to supply water and electricity to the arid Western regions. In hindsight, viewed through a lens in which people value free-flowing rivers, the ecological costs of America's large dams—hundreds of which dot the West—do in fact make this strategy look like a "mistake." But to many government officials, energy corporations, and even citizens in southwest China and Southeast Asia, dams still appear to hold the promise of cheap, reliable energy, not to mention revenue returns.

As we seek to understand the technical, financial, ecological, social, and even moral dimensions of hydropower development—and to avoid repeating many of the worst mistakes of the past—what role should scientists play? I recently sent a paper to a colleague for a friendly review prior to submitting it to a journal. In the margin of the section that described my role in interdisciplinary modeling efforts to improve decision making, she scrawled, "Oh, you're one of those kinds of anthropologists." Her comments were meant as friendly criticism, but also as a reminder to reflect on the risks, challenges, and contradictions involved in engaged scholarship.

I have chosen to participate in the highly charged discussions and debates surrounding the topic of dam construction for several reasons. First, I consider the basic question of how dams affect ecosystems and communities to be an empirical one. It should be approached with data, wise analysis, and critical discussion. Second, as my use of the moral economy framework is meant to emphasize, what we see often depends on what we include—and, just as crucially, what we exclude—in the analysis. If we consider the question of a dam's impacts to be an empirical one, then it follows that decision making can be done well by accounting for the fullest possible range of costs and benefits, or it can be done poorly by ignoring inconvenient data or by inviting only certain constituent groups to the table.

I appreciate the critical pieces of scholarship that highlight the myriad problems of large-scale development projects such as dams, but I don't think they end the conversation. Anthropologists and other social scientists tend to avoid any sort of prescriptive statements and often show a reflexive resistance to social change. But sociopolitical systems are complex, and we are often forced to live and work within ones that we don't find fully satisfactory. Whether we like it or not, the current question in China and in many other parts of the globe is not *whether* to build a dam, but rather how to build it, where to build it, and how to responsibly address the most egregious environmental and social ills through the use of mitigation measures or compensation programs.

Do we stand outside the system and critique it, or do we roll up our sleeves and engage with it? The costs of engagement, I have suggested, include the risk of being wrong, the risk of being labeled a reductionist by one's peers, and even the risk of embracing a fundamentally flawed system. But disengagement also comes with a set of costs, whether we acknowledge them or not. In short, disengagement allows for the perpetuation of the status quo, and the status quo is one in which hydropower-development corporations, armed with huge amounts of capital and increasingly vested with foreign shareholder money, can speak in a very loud voice to influence policy.

Participating in the discussion can help to promote development strategies that address the most important concerns for local people. The evidence presented on the Lancang River in this book shows that compensation measures are improving over time and that people facing resettlement in the near future can expect to fair better than their counterparts in the recent past. These steady improvements are partly the result of the tireless efforts by Chinese and foreign social scientists, journalists, and others who have worked to document the social costs of dams, to publicize their findings, and to advocate for better outcomes, often at considerable risk to their careers and their reputations.

THE RISE AND FALL—AND RISE?—OF HYDROPOWER

I have alluded to the long history of dam construction in many parts of the world, which stretches back millennia into the past. Dams continue to be viewed today as viable drivers of economic development in China and other emerging economies because of the electricity and revenue they generate. As I have shown in this book, the funding mechanisms and many other features of hydropower development have changed, but dams have proven remarkably durable as a development strategy.

By the 1970s, most industrialized countries, under pressure from the bourgeoning environmental movement, had largely stopped building hydropower dams—at least within their own borders. The commissioner of the U.S. Bureau of Reclamation stated with some confidence in 1994 that "the dam building era in the United States is now over" (qtd. in Worldwatch Institute 2001:37). Such a statement underscores the dramatic shift in public values in the United States, where most citizens and political leaders view hydropower generation as merely one positive outcome that must be weighed against a host of environmental and social ills—species loss, biodiversity decline, water-quality degradation, and human displacement. In many circles, hydrologists and environmental scientists call for managing the operations of existing dams for "ecological flows," an approach that balances human needs against ecological concerns such as fish passage, water quality, and the maintenance of healthy riparian habitats downstream.

But the eulogy for hydropower was perhaps a bit premature. Current events may yet shift the cost–benefit equation. Climate change, for example, may increase flood risks in some areas while depleting the natural storage capacity of snowpack in other areas (Nolin and Daly 2006); as a result, dams may again become a viable water-management tool in places such as the United States, which has not seen the development of additional hydropower facilities for many decades. Meanwhile, corporate interests and government factions in the United States are pursuing every source of hydrocarbons they can get their hands on. The Keystone XL pipeline, for example, which would transport crude oil from the tar sands of Alberta, Canada, to refineries in the Midwest and on the Gulf Coast, is in the late stages of review and permitting. Following the release of a draft EIA in 2013, James Hansen, then the director of the NASA Goddard Institute for Space Studies and the most well-known scientist in the public debate on global climate change, remarked somberly, "The total carbon in tar sands exceeds that in all oil burned in human history" (qtd. in Environmental News Service 2013). The extraction of crude oil from tar sands represents the turn toward dirtier fuel sources as the world's proven reserves of petroleum dwindle in our current era of "peak oil."[2]

All of this is taking place, moreover, amidst growing scientific consensus that average global surface temperatures are rising above historical norms, that the warming trend is correlated with anthropogenic carbon emissions, and that catastrophic climatic and weather events—which have become all too commonplace in recent years—are the likely result (Intergovernmental Panel on Climate Change 2007; Marcott et al. 2013). Water scenarios under climate change project retreating glaciers in some areas and less pronounced effects in others, which leads to questions about the future availability of freshwater in East, Southeast, and South Asian regions—home to about one-quarter of the world's population (U.S. National Research Council 2012). What are the implications of reduced snow pack in the Himalayas, often referred to as the "third pole" of the globe because of the sheer volume of freshwater stored there?[3]

At the international scale, various policy initiatives continue to make hydropower an attractive proposition for many governments as well as for private development interests. The international Clean Development Mechanism (CDM), for example, provides a means for carbon-emissions-reduction projects in developing countries. The CDM was devised to offer developing countries a measure of flexibility in meeting their obligations under the Kyoto Protocol. Developed countries (Annex I countries under the Kyoto Protocol designation) can earn Certified Emissions Reductions (commonly called "carbon credits") by investing in sustainable-development projects in the developing world. By showing that their financial assistance was key to setting up an emissions-reduction project that otherwise would not have gone forward for lack of financing, developed countries can subtract these carbon credits from their own portfolios as a way of meeting their obligations under the Kyoto Protocol.

A wide range of projects throughout the developing world, from biogas facilities to wind farms, receive foreign investment under the CDM. By a large margin, however, the category of projects that attracts the most CDM investment is hydropower (Thomas, Dargusch, and Griffiths 2011); corporations from around the globe see the potential financial benefits of investing in dams and counting such investments as part of their "green portfolio." Hundreds of dam projects throughout the world are under way with CDM assistance.

In a conversation with a staff member from CDM Watch, an NGO that monitors the carbon-exchange market, I learned that there are several problems with the CDM as it is currently conceived. First, the notion that dams result in a net reduction of carbon emissions is a questionable scientific proposition; whether hydropower is seen as "clean" or "green" depends ultimately on how one measures it. If it means reducing fossil-fuel consumption and greenhouse gas emissions, hydropower probably is clean, but we also have to acknowledge that large reservoirs in heavily vegetated areas usually release thousands or millions of tons of methane, one of the most potent of greenhouse gases, as the biomatter beneath the reservoir decomposes. If we include in this calculation the impacts on water quality, sediment, river morphology, riparian habitats and species, and other factors, hydropower is much less green than it first appears.

Perhaps the most problematic aspect of carbon-market schemes such as CDM relates to what policy documents refer to as "additionality." In order to receive carbon credits, an agency or company from an Annex I country must prove that a given emissions-reduction project—such as a dam—in a developing country would not have gone forward without financial assistance and that reductions in greenhouse gas emissions are therefore "additional." CDM Watch contends that the carbon-exchange market is rife with corruption, that powerful governments and corporate interests manipulate the system in their favor, and that decisions are made to move forward with many projects before a comprehensive analysis can be done.[4]

Meanwhile, global environmental discourse in recent years has largely shifted away from biodiversity conservation and related concepts and toward climate change as the principal concern. In the process, many hydropower-development interests seem more than happy to reframe their work in environmentally friendly terms as projects with the potential to reduce dependence on fossil fuels (Frey and Linke 2002). Moreover, in the absence of a truly game-changing technological advancement in the arena of energy production, energy policy often looks like a choice between bad options. As I have suggested, China's role in overseas dam building—through technical consulting and financing, through construction, and through the promotion of creative rights-transfer agreements—has become a major part of this story. On the subject of the global distribution of environmental ills, it is often said half-jokingly that "China needs a China." This clever turn of phrase refers to the fact that the country, which for a generation has been the dumping ground for dirty industries, now needs someplace to put the negative externalities of its own fast-paced development trajectory. In the huge expansion of dam building in Southeast Asia and sub-Saharan Africa, China appears to have found a China.

In the Chinese and international press, obituaries are written almost daily for China's great rivers. Dai Qing, the well-known environmental activist and one of the most outspoken critics of the Three Gorges Project from its earliest stages, commented that "the government built a dam but destroyed a river" (qtd. in Watts 2011). Even high-ranking party officials have begun to acknowledge to the media the fact that megaprojects such as the Three Gorges Dam face urgent problems, including water-quality deterioration, sediment buildup, and even joblessness and conflict in resettled communities. Of course, such problems would have been best considered before 16 million tons of concrete were poured across the Yangtze.

I wish to make an equally important, if somewhat less dire, appeal. If the dam projects in Yunnan continue to move forward, as it appears they will, there should be a public accounting of the costs, a solid understanding of who bears them, and clear mechanisms in place to pay them. Viewed in hindsight a generation from now, this period will be seen as a crucial turning point at which the world's most populous nation made critical decisions about how to power its future economic development. The measure of its success will be whether it can do so with a full accounting of ecological and social costs, an explicit acknowledgment of the trade-offs, and a commitment to greater transparency in decision making.

LIST OF CHINESE TERMS

Bulang zu	布朗族
chaiqian	拆迁
Chama Gudao	茶马古道
chengbao tian	承包田
chi, chuan, zhu	吃，穿，住
chunfen	春分
dagong	打工
Dai zu	傣族
da tiaozheng	大调整
Da Ziran Baohu Xiehui	大自然保护协会
di	地
ditan jingji	低炭经济
dongchong xiacao	冬虫夏草
dui shang bao gan, dui xia shi shi qiu shi	对上包干，对下实事求是
Dulong zu	独龙族
fa lü	法律
feizhengfu zuzhi	非政府组织
fubai	腐败
fupin wuqi	扶贫武器
gongping	公平
gongtong guanxin	共同关心
guai	怪
guanxi	关系
guojia youxian	国家优先

Han zu	汉族
Hui zu	回族
jianqing	减轻
jiazhi	价值
jie gaizi	揭盖子
Jihua Shengyu	计划生育
jin	斤
jinbu	进步
Jinsha Jiang	金沙江
jiti suoyou zhi	集体所有制
kaohe zhidu	考核制度
ke ai	可爱
kexue	科学
Lahu zu	拉祜族
Lancang Jiang	澜沧江
laobaixing	老百姓
Lisu zu	傈僳族
liuyu zonghe guanli	流域综合管理
liyi	利益
liyi xiangguanzhe	利益相关着
long sheng jiu zi	龙生九子
lüse fazhan	绿色发展
lüse nengyuan	绿色能源
meipo	媒婆
meiyou banfa	没有办法
minzu cun	民族村
minzu shibie	民族识别
minzu wenti	民族问题
mu	亩
Nan Shui Bei Diao Gongcheng	南水北调工程
Naxi zu	纳西族
neibu	内部
nongye shichanghua	农业市场化
Nu Jiang	怒江
Nu zu	怒族
pinggu baogao	评估报告
pinkun hu	贫困户

pinkun xian	贫困县
Putonghua	普通话
qingjie fazhan	清洁发展
qingjie nengyuan	清洁能源
qiufen	秋分
quan liuyu gong jihua	全流域共计划
quanli	权利
quanmin suoyou zhi	全民所有制
quntixing shijian	群体性事件
ren	人
ren zai guanyin yanghe zhong	人在观音养和中
san da	三大
San Tiao Hong Xian	三条红线
san tongshi	三同时
Sanjiang Bingliu	三江并流
shangfang	上访
shang ren chu, huang he qing	上人出，黄河清
shaoshu minzu	少数民族
shehui fengxian pinggu	社会风险评估
shehui jinbu	社会进步
shehui tuanti	社会团体
shenfen zheng	身份证
shengtai wenhua	生态文化
shuidian jidi	水电基地
shuidian kechixu fazhan	水电可持续发展
shuili shuidian	水利水电
Shuili Weiyuanhui	水利委员会
songbang	松绑
tao shuofa	讨说法
Tao Te Ching	道德经
tian	天 or 田
tongyi duo minzu guojia	统一多民族国家
tuigeng huanlin	退耕还林
Wa zu	佤族
wang po mai gua, zi mai zi kua	王婆卖瓜，自卖自夸

weiji	危机
wenhua	文化
wenze zhi	问责制
Wu Da Fadian Jutou	五大发电巨头
wuwei	无为
xiao zu	小组
Xibu Da Kaifa	西部大开发
Xi Dian Dong Song	西电东送
Xingzheng Fa	行政法
Yalong Jiang	雅砻江
Yao zu	瑶族
Yi Guo, Liang Zhi	一国两制
Yi zu	彝族
yiliu wenti	遗留问题
yimin cun	移民村
yuanzhu minzu	原住民族
zeren	责任
zeren tian	责任田
zhao qingtian	找青天
zhongguo tese de shehui zhuyi	中国特色的社会主义
zhuanjia	专家 or砖家
zige zhengshu	资格证书
zizhi qu	自治区
zizhi xian	自治县
zizhi zhou	自治州

NOTES

1. THE MORAL ECONOMY OF WATER AND POWER

1. This is a story that is global in scope. In Taiwan, hydropower potential is much smaller but is nevertheless an important part of the nation's energy mix. Hydropower development in Taiwan also poses some unique challenges, including threats to biodiversity, scenic areas, and sites of cultural significance for the nation's indigenous peoples (Taiban 2006). As I write, the news is filled with headlines about the massive Belo Monte Dam under construction on the Xingu River in the Brazilian Amazon; many news stories have featured photographs of Kayapo and other indigenous people being forcibly removed from the area.

2. Some scholars refer to Zomia as the "Southeast Asian Massif." I prefer the terms *highlands* and *uplands* to *massif* because a massif denotes an upthrust area of limited geographical scope, and Zomia encompasses highland areas from southern Vietnam to eastern Tibet, a vast region.

3. I don't want to lapse into the teleological language common among people who spend a lot of time thinking about "the state," which ascribes volition to an entity that has none. Even James Scott, whose book *Seeing Like a State* is perhaps the most far-reaching analysis of how states operate in the arena of economic development, cautions scholars against attributing too much agency to a governmental entity: "In trying to make a strong, paradigmatic case, I realize that I have risked displaying the hubris of which high-modernity is justly accused" (1998:7).

4. Michael Hathaway (2013, 2010) notes that some Yunnan-based scholars and NGOs—including the Center for Biodiversity and Indigenous Knowledge in Kunming—have begun strategically using the term *indigenous* to refer to local minority nationalities. They link indigenous status to the bourgeoning environmental movement and to struggles over land rights. A similar shift has already occurred in Taiwan, where native groups that once referred to themselves as "aboriginals" now commonly use the term *indigenous peoples* (*yuanzhu minzu*),

which puts them in line with the UN Declaration on the Rights of Indigenous Peoples and other international protocols.

5. Although literacy in Chinese is still a primary criterion of "culture" (*wenhua*) and "civilization" (*wenming*), various southwestern *minzu*, including the Yi, Dai, and Naxi, have had orthographies of their own for centuries.

6. The underlying logic of the "Er Hu" argument seems to be that socioeconomic disparities between *minzu* and Han Chinese have been solved largely through targeted economic assistance, preferential admissions policies in higher education, and related programs. In a recent seminar at the Minzu University of China, one senior scholar noted, with more than a small dose of satire, that minority issues in China have been "solved" in the same sense that racism in the United States ended when an African American man was elected president.

7. Immanuel Kant, perhaps the most influential moral philosopher of all time, argued strongly that for morality to have any force at all, it must exist in a universal form. This argument is problematic for anthropologists, for whom cultural relativism has served as an ideological keystone over the past half-century. Like most anthropologists, I am interested less in abstract questions about what is "moral" than I am in figuring out how morality is constructed and maintained through social, cultural, and political practices. Jarrett Zigon (2008) provides a thorough treatment of morality in anthropological and cross-cultural perspective.

8. Of course, concentrated political power can occasionally be wielded for benevolent purposes. While the United States remains mired in political bantering about what role the government should play in promoting "green" technologies and jobs, China is moving forward with major initiatives to develop energy from solar, wind, and other renewable sources. At a sustainability fair in Oregon, where I live, a group of engineers from a small start-up company showed off their "new" invention: a passive solar water heater that warms water in a series of polyvinyl tubes without the need for a photovoltaic panel, stores the water in a container, and releases it on demand for household use. The irony is that this relatively simple invention has been in use in tens of millions of homes and businesses in urban and rural China for the past decade. Even the poorest households in rural Yunnan have them. Their ubiquity is evidence of sound policy promotion: the NDRC has sponsored a campaign to promote such devices, and the central government has provided subsidies for poorer regions and communities.

9. Indeed, the Jinsha, the headwaters of the Yangtze, is undergoing a massive hydropower-development effort. In a recent news article titled "China's Five Hydropower Giants Divide Up the Jinsha River: A Reservoir Every One Hundred Kilometers," the *People's Daily* (2012) reported that the river's 2,308 kilometers may eventually be home to as many as twenty-five hydropower dams. Environmental organizations are worried that the Jinsha will essentially be turned into segment after segment of still water.

10. The list of Western adventurers who made their way through Yunnan in the nineteenth and early twentieth centuries is long: the French Catholic priest Pere Paul Vial; the American Baptist missionary David Crockett Graham; the Austrian-born American naturalist Joseph Rock; the British botanists George Forrest and Francis Kingdon Ward; and many others. The collected volume *Explorers and Scientists in China's Borderlands, 1880–1950*, edited by Denise Glover and her colleagues (2011), provides the best treatment to date of this motley group, the motivations that drove them, and the intellectual fruits that were born of their labors, including seed and specimen collections, religious conversions, minority-language dictionaries, and beautiful photographic records of landscapes and people.

2. CRISIS AND OPPORTUNITY: WATER RESOURCES AND DAMS IN CONTEMPORARY CHINA

1. The global financial crisis that commenced in 2008 poses a challenge to the CCP, which must cope with possible social unrest from slowed economic growth and rising unemployment. Still, the Chinese economy has fared much better than the European and the U.S. economies; newspapers reported that overall economic expansion had slowed to "only" 8 percent by the first quarter of 2012, a level of growth most other countries could not hope to achieve.

2. The life-cycle approach is the best way to understand the total carbon intensity of different electricity sources, but it is not without complications. The study cited here is based on the review of dozens of studies conducted mostly in the United States and other highly developed economies. Comparable studies on China are hard to come by; given differences in technology and scale, Chinese figures are likely to be similar but not identical to the figures provided here.

3. One bright spot is the gains in energy efficiency in recent years, particularly in heavy industries such as iron and steel, driven by capital investment in cleaner technologies (see Managi and Kaneko 2010). These technological advances are offset by the fact that economic globalization makes it "rational" for foreign corporations to locate their dirtiest production facilities in China.

4. A recent meme about the hazy skies has been popular on social media Internet sites in Beijing: couples walking down the street are said to remark, "I'm holding your hand, but I can't see your face" (*Wo qianzhe ni de shou, kan bu dao ni de lian*).

5. Even good news such as the Chinese expansion of renewable energy production is often met with menacing headlines such as "China Seeks Dominance in Clean Energy" (Bradsher 2009). Western governments complain about China's "green protectionism," citing requirements that most equipment for solar panels and wind turbines be made by Chinese manufacturers.

6. Premier Wen Jiabao walked away from the Copenhagen talks over the issue of whether developing nations should be held to the same levels of emissions reductions as developed countries. A key sticking point was exactly how to

calculate emissions: in the aggregate, which would rank China first in the world in emissions, or per capita, in which case China's emissions would be close to the global average.

7. Oil is economically feasible as a principal fuel source only as long as its availability and its price remain within a certain range. In theory, as soon as another energy source is capable of replacing oil at a competitive price, it will do so. Of course, oil remains the fuel of choice in part because many of its costs—pollution, toxic hazards, health threats, and climate change—either go unexamined or are externalized onto regions, nations, or individuals with less political and economic power.

8. The North China Plain offers a good illustration of what a "peak-water" scenario might look like. As water demand increases, projects designed to increase supply—including dams and reservoirs—expand. Once it is no longer cost effective to extract local surface water and groundwater, we would expect to see a shift to more costly measures such as desalinization or interbasin water transfers (Palaniappan and Gleick 2009:11). This is precisely what has happened in the North China Plain, where extraction of water from the aquifer has far outpaced the natural recharge rate. Desalinization of seawater is one option currently being pursued. The South–North Water-Transfer Project, which may yet become the largest engineering project undertaken anywhere in the world, is another option that is already under way, as I describe in this chapter.

9. The South–North Water-Transfer Project involves significant population displacement, although the precise figures are not well known. In May 2012, China Central Television channel 13 broadcast a report on displaced villagers as part of a regular segment called *Into the Grassroots* (*Zou Jiceng*). The report featured video footage of tearful residents leaving their hometown, a small village in Hubei Province northeast of Wuhan, as bulldozers and backhoes demolished their houses. But the report was careful to end on a positive note, showing villagers moving into newly constructed houses with government sponsorship.

10. One U.S. dollar is approximately 6.2 yuan.

11. Several *aflaj* systems in Oman were designated UNESCO World Heritage Sites in 2006. The ecological sensibility of these systems is impressive. Whereas modern pump wells can extract quantities of water capable of depleting aquifers and causing saltwater intrusion, *aflaj* systems work by gravity and rely upon aquifer equilibrium to get the water to flow where it is needed (see Abdulbaqi et al. 2007). Similar systems are known by different names throughout the Middle East, including *falaj, kahan,* and *qanat.* Although such systems are apt examples of the "soft-path" approach to water utilization, it is admittedly hard to see how they might address water-allocation problems involving huge populations.

12. At the other end of the flow-regulation spectrum, reservoir storage in the Amazon–Orinoco is only about 3 percent of the mean annual discharge. Hydropower development on the Amazon, particularly in Brazil, is gaining speed and momentum.

13. ICOLD was founded in 1928; since that time it has grown to include representatives from more than eighty countries. ICOLD's central mission is to disseminate information about best practices in the science and engineering of dams. In the early and mid–twentieth century, it was instrumental not only in the growth of dam construction throughout the world, but also in improving dam safety and reducing the rate of dam failures (see Schnitter 1994).

14. More recently, Chinese leaders have learned from the U.S. example about the consequences of rapid deregulation in the energy sector. During the California electricity crisis of 2000–2001, residents experienced electricity shortages and rolling blackouts as large energy companies, including Enron, manipulated market supplies to make huge short-term profits. This concrete lesson probably served to slow down the pace of reform in China (see Yeh and Lewis 2004:449).

15. Hydropower development in the Pacific Northwest was also intimately linked to national security and defense. The Columbia River supplied both water and power for the Hanford Nuclear Reservation, which manufactured plutonium for the "Fat Boy" bomb dropped on Nagasaki, Japan, in 1945 and for much of the U.S. nuclear arsenal during the Cold War. Today, the Hanford Nuclear Reservation, where nuclear waste is slowly leaking from underground storage containers, is the largest site of environmental remediation in the United States.

16. The Elwha Ecosystem Restoration Project, which involves the removal of two dams on the Elwha River, located on the scenic Olympic Peninsula in Washington State, is the largest effort to date. Allowing the passage of salmon upstream to spawning grounds is a major rationale cited by proponents of the project. Efforts like these serve to highlight the evolution of ecosystem science over the past century: when these dams were built, scientists had little knowledge of the important role migratory fish play in conveying marine nutrients inland into terrestrial habitats.

17. Dujiangyan (in Sichuan Province), Zhenguo Canal (in Shaanxi Province), and Lingqu Canal (in Guangxi Province) are considered the three great hydrological engineering projects of the Qin Dynasty. The Dujiangyan project has proved remarkably resilient. Following the 8.0-magnitude Wenchuan earthquake in 2008, which was centered only a few tens of kilometers away, one of the project's three main levees was slightly damaged, but the project was otherwise unscathed.

18. More than forty Chinese companies are listed among the Global Fortune 500; the majority are state-owned enterprises in key sectors such as petroleum and energy.

19. Twelve administrative units are targeted in the Great Western Opening campaign, including one municipality under central-government jurisdiction (Chongqing), five minority autonomous regions (Tibet Autonomous Region, Ningxia Hui Autonomous Region, Xinjiang Uighur Autonomous Region, Inner Mongolia Autonomous Region , and Guangxi Zhuang Autonomous Region), and six provinces (Sichuan, Guizhou, Yunnan, Shanxi, Gansu, and Qinghai).

3. THE LANCANG RIVER: COPING WITH RESETTLEMENT AND AGRICULTURAL CHANGE

1. DNA research has recently led to the reclassification of the caterpillar fungus. It is now considered *Ophiocordyceps sinensis*, although many people still refer to it as *Cordyceps*.

2. Toponyms can be one of the most confounding aspects of conducting research in this corner of southwest China, where dozens of languages, mostly Tibeto-Burman in origin, coexist. Khawa Karpo, the highest peak in Yunnan, is a case in point. Its name comes from a transliteration of its Tibetan name "Kawagebo," but it is commonly transliterated as "Kawa Garbo," "Kawagarbo," "Kawadgarbo," or "Kha-Kar-Po." Just to confuse matters further, the Chinese refer to the peak as "Meili Xueshan" (Meili Snow Mountain). At 6,740 meters, Khawa Karpo is the highest mountain in Yunnan.

3. Initial plans called for eight dams in the Lancang upper cascade. The Mengsong Dam was cancelled, although some experts speculate that it may be put back on the agenda.

4. The *mu* is a standard unit for measuring land area in China. One *mu* is equal to 0.066 hectares or 0.165 acres.

5. The Manwan Dam environmental and social impact study was published as *Master Report of the Study on Manwan Dam Related to the Social, Economic, and Environmental Impacts on the Lancang River*, which was jointly completed by He Daming, Yu Xiaogang, Chen Lihui, Guo Jiaji, Gan Shu, and Li Qin and translated by Oxfam Hong Kong from Chinese into English in 2002.

6. Reservoir size is the key to obtaining accurate estimates for population displacement but is often the subject of scientific dispute. For example, Kelly Kibler (2011) conducted a geospatial analysis of the dams in the Lower Cascade project on the Lancang. Based on location, design characteristics, and topography, she was able to estimate the spatial extent, or "footprint," of each reservoir. She found that some official estimates of reservoir sizes by Chinese government and hydropower corporations were far too low, often by as much as a factor of ten.

7. The New Rural Cooperative Medical Care System, initiated in 2003, was designed to improve health-care access among the rural poor. It is a tiered system that allows patients to access local village and township clinics and county hospitals, with heavy subsidies from the central and provincial governments. Along with other policy initiatives, especially the New Socialist Countryside program, it represents a major government effort to address the growing economic disparities between city and countryside.

8. On this particular trip, with the primary intention of pretesting our household survey, our research team inadvertently created a new sampling technique that we jokingly referred to as the "muffler sampling frame." We were trying to reach

two particular villages in Xiaowan Township from which we intended to sample, but about twenty minutes outside the township center, as we were rattling along dirt and cobble roads, our truck's muffler fell off and started dragging along the ground. The driver, a local man, spent several hours locating someone to fix it and another several hours actually fixing it, and we used that time for conducting some initial pretests of the survey.

9. After many years of working with hydrologists, geographers, and other spatially inclined people, I have come to visualize these study areas in terms of watersheds. When I present tabular data on the study areas or dam sites in this book, I do so from north to south, which, conveniently for both the Nu River and the Lancang River, means from upstream to downstream. Survey data come from the Integrative Dam Assessment Model Socioeconomic Survey (Brown and Tilt 2010).

10. Other researchers have come up with alternative solutions to the problem of understanding diachronic change due to resettlement. Michael Webber and Brooke McDonald (2004), for example, examined the effects of resettlement at Xiaolangdi Dam on China's Yellow River by asking resettled villagers retrospective questions about their socioeconomic status prior to the resettlement campaign. Although this approach has the advantage of measuring change between time 1 and time 2, it also introduces error because villagers are asked to recall crop yields, household income, and other information from years in the past.

11. Total household income was calculated in a way that goes beyond typical government accounting measures. The *Yunnan Statistical Yearbook* (Yunnan Statistical Bureau 2011) estimates average annual household income in these counties to be about 19,000 yuan (5,000 yuan per capita, a little more than U.S.$800, with an average household size of 3.8 persons). The figures reported in our socioeconomic survey are notably higher for two reasons. First, the survey calculated income in a much more comprehensive way, including agricultural sales minus the costs of inputs such as fertilizers and pesticides; the value of agricultural and livestock production for household consumption; wage labor; and self-employment. Second, these survey income figures also include government subsidies—both poverty-alleviation subsidies and, for many households, displacement compensation—which are addressed in detail in chapter 6.

12. Chi-square tests were used to compare proportions across the study groups, and independent-samples t-tests were used to compare means.

13. To avoid respondent bias and encourage study participants to consider their answers carefully, some questions were phrased negatively, and the responses were inverted during data analysis.

14. These days palm-fiber capes are seen more often in ethnological museums than as practical clothing. Villagers make them by stripping the fibrous sheathing from a palm tree trunk, flattening it, and weaving the strips together, then fastening them with a length of rope to hold them all together.

4. THE NU RIVER: ANTICIPATING DEVELOPMENT AND DISPLACEMENT

1. Joseph Rock and Francis Kingdon Ward, along with many other naturalists and explorers during the late nineteenth and early twentieth centuries, tended to view their work as objective classification of the biological and social world. Viewed in retrospect, their efforts seem positivist at best and colonial at worst. But there is no doubt that they were careful and comprehensive scholars—so comprehensive, in fact, that, as Stevan Harrell points out, Naxi people in recent years have reprinted Joseph Rock's dictionary of their language, which is Tibeto-Burman in origin and uses a unique set of pictographic characters, in an attempt to revitalize their linguistic heritage following repression under the CCP (2011:8). The repressive atmosphere, which reached its zenith during the Cultural Revolution (1966–1976), had a particularly harsh effect on the cultural practices of many of southwest China's ethnic minorities (see Mueggler 2001).

2. Marco Clark (2009), who conducted several months of fieldwork in the upper Nu basin in 2008, reports that more than half of the villages in Dimaluo Township had Catholic churches. A priest from the county town of Gongshan travels once each month to outlying villages to administer mass.

3. China's Planned-Births Program (Jihua Shengyu) strictly limits urban households to one child but provides less-stringent regulations for rural areas with large minority populations and areas where a significant percentage of residents live below the poverty level. (These two conditions, in fact, often coincide.) Families in Fugong County, most of whom rely on labor-intensive agriculture to make their living, are presently allowed two children.

4. Called the Fraser Alphabet, this is the official orthography of the Lisu language. James O. Fraser, a British Protestant with the China Inland Mission, pioneered the script by using the Latin alphabet and rotating the letters in various ways to represent consonants and vowels unique to Lisu, thus eliminating the need for diacritical marks. Like Chinese, Lisu is a tonal language, and the Fraser Alphabet uses common punctuation marks to convey tones. (See J. Goodman 2012:185.)

5. Jim Goodman notes that the modern steel-cable ziplines actually constitute a significant infrastructural improvement. Until the 1950s, most ziplines were fabricated out of interwoven strands of split bamboo. Moreover, the caravan route up the canyon, now a major two-lane road paved with asphalt, was previously carved out of the stone canyon walls; the most precarious sections of the route were supported by wooden scaffolding.

6. The predecessor of the State Electric Power Corporation was the Ministry of Electrical Power.

7. Similar to linseed oil, tung oil is used in wood finishing. Although the number of villagers who grow tung oil trees in the Nu River Gorge is small, it can be a lucrative enterprise.

8. As recently as 2009, Western and Chinese linguists "discovered" multiple languages in Yunnan—not dialects, but languages with fully formed grammatical systems and syntaxes that were mutually unintelligible. None had been previously described by scholars. (See Erard 2009.)

9. In Taiwan, by contrast, native groups who previously referred to themselves as "Aborigines" (*ben tu*) have more recently applied the term *indigenous peoples* (*yuanzhu minzu*) to themselves, likely wishing to gain any possible political advantage from the passage of the UN Declaration on the Rights of Indigenous Peoples. Significantly, Article 10 of the declaration stipulates that "indigenous peoples shall not be forcibly removed from their lands or territories" and that any relocation must take place with informed consent and with "just and fair compensation and, where possible, with the option of return."

10. Indeed, even responding to survey questions about the dam projects can be politically risky. Most villagers hesitate to overtly oppose the dams or to ally with antidam NGOs (Magee and McDonald 2009).

5. EXPERTS, ASSESSMENTS, AND MODELS: THE SCIENCE OF DECISION MAKING

1. This particular episode predates the emphasis on "ecosystem services" that is now common in environmental policy circles. Located at the intersection of ecology and economics, the ecosystem services concept suggests that one way of valuing the biophysical environment is to ask, "How much would it cost to replace services such as water filtration or erosion control currently provided by natural processes?"

2. The extent to which NEPA succeeds in integrating public views into decision making is debatable. Many environmental communications experts suggest that collaborative decision making can still be dominated by government agencies, jokingly referring to the process as the "three I's model": "inform, invite, ignore." The government agency informs the public of a pending action, invites them to participate in the collaborative decision-making process as required by various federal statutes, and then promptly ignores their input. (See Leong, Forester, and Decker 2009.)

3. There are some serious limitations to EIAs' scope of influence in the United States, two of which are worth mentioning briefly. First, certain projects can receive a categorical exclusion from the assessment process. For example, federal courts recently ruled that the U.S. Navy did not violate NEPA by continuing low-frequency sonar testing off the western coast, despite scientific findings that this technology harmed marine mammals. The courts found that an executive decision, based on matters of national security, superseded NEPA. Second, environmental management agencies in the United States are not required to choose the most "environmentally benign" management alternatives; rather, they simply

need to demonstrate that they considered all of the information in an EIA report during the process of reaching a decision.

4. Although Hong Kong has nominally been part of the PRC since reverting to Chinese control in 1997, it operates under a rubric called "One Country, Two Systems" (Yi Guo, Liang Zhi), in which Beijing minimally interferes in its political and economic affairs.

5. The IDAM project was geared toward developing a model with general applicability in China and elsewhere. Our Chinese partners from the Asia International Rivers Center at Yunnan University and from Yunnan Normal University provided valuable collaboration and intellectual input throughout the project.

6. One of the most far-reaching attempts to improve the environmental and social outcomes of dam construction is the Hydropower Sustainability Assessment Protocol, developed by the International Hydropower Association, a nonprofit organization based in London with support from the hydropower industry. Several engineers within China's Institute for Water Resources and Hydropower Research told me that they were actively using the association's protocol to assess current hydropower projects.

7. Over the years, I have heard a variety of critiques leveled at modeling projects such as the IDAM. Such critiques tend to come from all parts of the theoretical spectrum: some natural scientists have claimed that the model is too general, while some social scientists have rejected the modeling enterprise as altogether too reductionist and positivist. My response is usually to point out how extensively we use models in our everyday lives—from weather reports to metaphor in poetry. After all, when Shakespeare penned his sonnet 18—"Shall I compare thee to a summer's day?"—I doubt that anyone rebuffed him by pointing out that his lover was not, in fact, a summer's day.

8. Data-modeling projects in science have exploded over the past few decades, reflecting the exponential growth of computing power available to experts and laypeople alike. Our group's modeling efforts are but one small corner of this vast field. In 2009, the Technical Advisory Committee for Environmental Research and Education at the U.S. National Science Foundation (2009) released a report on how advancements in computer science are changing the face of environmental research. In some cases, data collection and storage are far outpacing scientists' ability to perform the work of analysis and interpretation.

9. The World Wide Fund for Nature (2004) has identified "cumulative impacts" as one of its central concerns over the Nu River projects, given the high biodiversity of the region and the previously undammed status of the river.

10. Kelly Kibler (2011) conducted an extensive comparison of small- and large-scale hydropower projects as part of her doctoral dissertation. She found that a series of small dams—which environmental groups often see as a favorable alternative to large dams—can actually have negative effects that surpass those of large dams. Part of the reason lies in the regulatory structure. Whereas large dams

must go through a complete EIA process, small dams can often be approved at the county level. More crucially, because local governments tend to face revenue shortages, they often approve small-scale hydropower projects out of economic exigency. (See also McDonald 2007.)

11. Beyond mitigation of potential impacts, many scholars and advocates increasingly call for reparations for past injustices related to large-scale development projects. (See, for example, Johnston 2012.)

6. PEOPLE IN THE WAY: RESETTLEMENT IN POLICY AND PRACTICE

1. The problematic distribution of costs and benefits related to dams is hardly unique to China. One particularly egregious case is the Lesotho Highlands Water Project, where a series of dams and tunnels have been constructed to deliver water from Lesotho to South Africa. Embraced by Lesotho's national leaders for its contribution to the national economy, including major investments from the World Bank and the African Development Bank, the project has brought resettlement, health problems, and even reservoir-induced seismicity to local residents. On this project, see, for example, Tilt, Braun, and He 2009.

2. If public opposition is difficult to characterize, so, too, are the outcomes associated with opposition. In his book *China's Water Warriors*, the political scientist Andrew Mertha (2008) provides three case studies of public opposition to dams: Dujiangyan, in Sichuan Province, where plans were modified to avoid damaging the 2,200-year-old cultural heritage site; Pubugou, also in Sichuan, which generated public protest that was widespread but failed to precipitate change; and the Nu River dams in Yunnan, which are described in detail in this book.

3. Further incentive for land conversions stems from the nationwide abolition of the agricultural tax in 2006. Although most farming households welcomed the removal of this tax burden, it has also caused local government officials to pursue nonagricultural land uses with greater zeal in order to generate revenue.

4. The term *chaiqian* (literally "demolishing and relocating") is used to describe land requisition for urban redevelopment. It is one of the most controversial social issues in China today, involving thousands of protests each year, some of which turn violent. The controversy surrounding *chaiqian* stems in part from the ambiguity of current property law in China and the competing interests of government agencies, rural collectives, citizens, and private development companies (see Ma Jianbo 2013; Zhang 2010). In some cases, villagers give up their land rights in exchange for a "residential household registration" (*jumin hukou*), which allows them to live in urban centers. Meanwhile, private development companies and local governments, who "stir-fry the land" (*chao di*) by converting agricultural fields to residential or commercial property, can make huge sums of money.

5. The idea of revisiting past compensation is in line with current international trends. Many scholars now argue that mitigation of present damages is not enough and that reparations for lost livelihoods due to displacement should be made. A case in point is the Chixoy Dam in Guatemala, built in the 1970s and 1980s with financial backing from the World Bank and the Inter-American Development Bank and conspicuously lacking a comprehensive assessment of social impacts or a resettlement plan. Thousands of people, mostly Maya Achi from Río Negro and nearby communities, were displaced, permanently losing access to farmland, grazing lands, forest resources, and fisheries. When locals put up resistance, they were met with military force (Guatemala was in the midst of a civil war), and an estimated 5,000 people were killed in what became known as the Río Negro Massacres. Decades of research provided well-documented evidence, presented to the Guatemalan government and the UN special rapporteur on the rights of indigenous peoples, that the Chixoy Dam was the proximate cause for such misery. The Guatemalan government agreed in 2010 to provide reparations to the communities, but the agreement has as yet not produced any concrete reparations (see Johnston 2012).

6. Recent scholarship has outlined the considerable evolution in China's compensation-policy framework (Wang et al. 2013). In line with policy changes since the late 1990s, the central government has retroactively increased compensation levels for some resettled villagers at Manwan and other sites.

7. The Five Energy Giants have undoubtedly noted the increasingly generous compensation policies. There is evidence to suggest that, faced with the possibility that compensation requirements will be further raised, these companies are pursuing a strategy of "preemptive development," seeking to build more dams as quickly as possible before compensation payments further cut into their profits (*People's Daily* 2012).

7. A BROADER CONFLUENCE: CONSERVATION INITIATIVES AND CHINA'S GLOBAL DAM INDUSTRY

1. In China, the panda is undoubtedly such a symbol. It has been used as the mascot for the World Wide Fund for Nature (formerly the World Wildlife Fund) since the organization was founded in 1961. There is substantial uncertainty about the total number of species in the world, with estimates ranging from a few million to more than 50 million. Only a small portion of the world's species has been well described scientifically. Vertebrates and other species that are closer to humans on the evolutionary chain are more likely to be well known, whereas only about 2 percent of extant nematodes (microscopic worms) are well described.

2. At a more fundamental level, nature conservation in China has required the importation of such terms as *nature* (*ziran*) and *environment* (*huanjing*) from the Western ecological sciences. This cross-cultural borrowing often took place via

Japan, which opened its doors to Western science and commerce decades before China. Thus, Western scientific terms effectively displaced traditional Chinese semantic categories for nature such as *tian* (literally "heaven") and *shanshui* (literally "mountains and water"). (See Weller 2006:43.)

3. The fact that interpretive materials at Pudacuo National Park contain little information about the region's human inhabitants may seem unremarkable. After all, most local residents are Tibetan, a group whose political subordination remains a national sore point. But many of China's other most vaunted natural sites have chosen to integrate culture into their activities and programs. In Guilin, for example, tourists typically take a Li River cruise to view the region's famous karst formations—the craggy mountains featured in many classic landscape paintings—but they also tend to visit one of the many local ethnic minority villages (*minzu cun*) that showcase the traditional costumes, songs, and dances of groups such as the Zhuang and Yao.

4. For an extensive overview of how Chinese scholars and activists working in the arenas of environmental protection and indigenous rights have used ties with Western organizations to accomplish their goals, see Hathaway 2010 and 2013.

5. China now has two activists in the water-resource sector who have received the prestigious Goldman Prize. In 2012, Ma Jun was awarded the prize for his decades of work as a journalist and activist uncovering violations of environmental quality. His book *China's Water Crisis*, published in 1999, was one of the earliest and most comprehensive scholarly treatments of China's environmental problems.

6. Research in India, where hydropower development is proceeding at a similarly rapid pace, suggests that the same is true there. Analysis of the Sardar Sarovar project in the western state of Gujarat indicates that the most successful advocacy movements tend to find ways to build bridges between local communities and international constituents, including NGOs, scholars, and the media. (See Oliver-Smith 2010; Bavishkar 2005.)

7. Critics of the excesses of conservation-minded Big International NGOs such as TNC rightly point out that nature conservation can have deleterious consequences for people. In particular, local and indigenous communities whose land is targeted as a protected area routinely find themselves without access to grazing land, farmland, firewood, hunting, and other resources that were once integral to their livelihood strategies. (See Dowie 2009.)

8. Five dams are planned for the downstream reach of the Mekong in Myanmar, including the controversial Tasang Dam, which at more than 185 meters would be the tallest dam in mainland Southeast Asia. Initial financing has been provided by Japanese government agencies and corporations, with technical and engineering assistance from Sinohydro Corporation and the China Southern Power Grid Company.

9. The MRC's effectiveness is also routinely undercut by bilateral arrangements that produce favorable financial outcomes for one or more partner countries.

Darrin Magee (2011) points out that several of the Lancang dams were the beneficiaries of investment by the Electricity Generating Authority of Thailand beginning in the 1990s.

10. Similar to the Lancang case, international relations in the Nu/Salween basin are
under increasing stress. The Transboundary Freshwater Dispute Database cata-
logs events related to water governance around the world, using media reports,
government documents, and other sources, and a group of researchers rates
each event on a scale from "cooperative" to "conflictive." An analysis of this database for the Nu River shows increasingly conflictive relations over the past two
decades (see Tullos et al. 2013).

11. As of 2003, World Bank officials made a considered decision to get back into dam
development. According to a senior official in the bank's Africa Program, this
decision was born of the fact that bank officials had recalibrated their thinking
about how to balance "sins of commission" (e.g., the ecological and social costs
of dams) against "sins of omission" (e.g., failing to fund projects that, undertaken
well, have the potential to spur economic development and alleviate poverty).
In 2012, for example, the World Bank announced its intention to partially fund,
along with various other international financial institutions, the Lom Pangar
Hydropower Project in Cameroon, which is expected to cost nearly $500 million
(Reuters 2012).

12. As I outline in the concluding chapter, this is precisely what has occurred. The
CDM, under the Kyoto Protocol, allows companies from developed countries to
invest in energy-generating projects, including dams, in the developing world
and to receive emissions-reduction credits for doing so.

13. The Chinese dam industry's involvement overseas is increasing exponentially.
A report by International Rivers in 2012 estimated a total of 308 dam projects in
seventy countries with significant involvement by Chinese companies or finan-
ciers.

14. The political landscape of Myanmar has changed dramatically in recent months
and years. Democratic elections were held in 2012 under international supervision, and the longtime political dissident and opposition leader Aung San Suy Kyi
won a seat in Parliament.

15. Another looming geopolitical wrangle involves hydropower development on the
Yarlung Tsangpo River, which flows from west to east along the northern edge
of the Himalayan Range before heading south, where it is known in India as
the Brahmaputra. Chinese companies are currently working to build at least five
dams on the upper segment of the river in the Tibet Autonomous Region, causing concerns in India about water security and prompting governmental support
for Indian state-owned companies to construct their own projects in the state of
Arunachal Pradesh. (See Hennig et al. 2013.) The two sides signed a memoran-
dum of understanding in 2013, which facilitates the sharing of hydrological data
and may be a first step toward more collaborative relations (Narayan 2013).

CONCLUSION: THE MORAL ECONOMY REVISITED

1. Improvements in dam operation for "environmental flows" in the United States owe a great debt both to shifting public values and to legal statutes such as the Endangered Species Act of 1973. In a conversation with a scientist from the Bonneville Power Administration, which markets the electricity produced by dams in the Pacific Northwest, I learned that the organization takes in approximately $3 billion in revenue each year and spends nearly one-third of that figure on scientific studies, technological innovation, and infrastructural improvement to aid fish stocks, especially Pacific salmon.

2. Hydraulic fracturing represents another blind commitment to fossil fuels. "Fracking" to extract natural gas is a process with severe environmental impacts and a poorly conceived regulatory structure.

3. I visited Jade Dragon Snow Mountain (Yulong Xueshan) with my family in 2012. We spent the afternoon on the mountain's glacier, which, at 4,500 to 5,500 meters in elevation, towers above Tiger-Leaping Gorge, its melt water feeding the Jinsha River. Like glacial peaks throughout the eastern extent of the Himalaya and beyond, there is concern here that climate change will cause glacial retreat, threatening the water supply of millions downstream.

4. Even official reports from the European Union and the PRC on their joint efforts to combat climate change reflect a sense of optimism mixed with caution when it comes to the CDM. Although both sides report a "common concern" (*gongtong guanxin*) over the effects of climate change and a shared commitment to find policy solutions, they also admit to fundamental problems such as the difficulty of certifying emissions reductions and the challenge of carrying out technology-transfer programs. (See China–EU Clean Development 2012.)

WORKS CITED

Abdulbaqi, Al Kabouri, Muyibi A. Suleyman, Ahmed M. Thamer, and Kabbashi A. Nassereldeen. 2007. "Integrated Water Resource Management Plan (IWRMP) in Oman: Way Forward." *Water Policy* 9:457–473.

Atwill, David G. 2005. *The Chinese Sultanate: Islam, Ethnicity, and the Panthay Rebellion in Southwest China*. Stanford: Stanford University Press.

Baviskar, Amita. 2005. *In the Belly of the River: Tribal Conflicts over Development in the Narmada Valley*. 2nd ed. Oxford: Oxford University Press.

Bodley, John. 2008. *Victims of Progress*. 5th ed. London: AltaMira Press.

Boekhorst, D. G. J., T. J. M. Smits, X. Yu, L. Li, G. Lei, and C. Zhang. 2010. "Implementing Integrated River Basin Management in China." *Ecology and Society* 15(2): 23–31.

Bonheur, N. and D. B. Lane. 2002. "Natural Resources Management for Human Security in Cambodia's Tonle Sap Biosphere Reserve." *Environmental Science and Policy* 5:33–41.

Bourdieu, Pierre. 2004. "The Peasant and His Body." *Ethnography* 5(4): 579–599.

——. 1990. *The Logic of Practice*. London: Polity Press.

Boyd, Olivia. 2012. "China Brings Dams Back to Africa." *The Third Pole*, July 10. http://www.chinadialogue.net/article/show/single/en/5032-China-brings-dams-back-to-Africa.

Bradsher, Keith. 2012. "'Social Risk' Test Ordered by China for Big Projects." *New York Times*, November 12. http://www.nytimes.com/2012/11/13/world/asia/china-mandates-social-risk-reviews-for-big-projects.html?_r=1&.

——. 2009. "China Seeks Dominance in Clean Energy." *New York Times*, July 14.

Brockington, Dan. 2002. *Fortress Conservation: The Preservation of the Mkomazi Game Reserve*. Bloomington: Indiana University Press.

Brody, Liz. 2012. "MEP Urges Greener Hydropower Development." *China Green News*. http://eng.greensos.cn/ShowArticle.aspx?articleId=1276. Accessed May 10, 2013.

Brown, Philip H. and Bryan Tilt. 2010. "Lancang River Socioeconomic Survey Data Set." Unpublished data set, Corvallis, Oregon.

———. 2009. "Nu River Socioeconomic Survey Data Set." Unpublished data set, Corvallis, Oregon.

Brown, Philip H. and Yilin Xu. 2010. "Hydropower Development and Resettlement Policy on China's Nu River." *Journal of Contemporary China* 66(19): 777–779.

Cain, Nicholas L. and Peter H. Gleick. 2005. "Real Numbers: The Global Water Crisis." *Issues in Science and Technology* 21(4): 79–81.

Cashmore, Matthew. 2003. "The Role of Science in Environmental Impact Assessment: Process and Procedure Versus Purpose in the Development of Theory." *Environmental Impact Assessment Review* 24:403–426.

Cernea, Michael. 2000. "Risks, Safeguards, and Reconstruction: A Model for Population Displacement and Resettlement." In Michael Cernea and Christopher McDowell, eds., *Risks and Reconstruction: Experience of Resettlers and Refugees*, 11–55. Washington, D.C.: World Bank.

Chen, Hongwei. 2006. "Nujiang shuidian kaifa 'da tiaozheng' fang'an wu xiawen: Baixing bei meng zai guli" (The Nujiang hydroelectricity-development 'adjustment': People were kept in the dark). *China Economic Times*, June 21. http://finance.sina.com.cn/roll/20060621/0902756782.shtml.

China Daily. 2011. "Nujiang Hydro Project Back on Agenda." Beijing, February 1.

———. 2010. "China Denies Dams Worsen Drought in Mekong Basin." Beijing, March 31. http://www.chinadaily.com.cn/china/2010-03/31/content_9664697.htm.

———. 2009. "Ministry Suspends Key Projects." Beijing, June 12. http://www.china.org.cn/business/2009-06/12/content_17934702.htm.

China–EU Clean Development Mechanism Program. 2012. *2012 Nian Qian Zhongguo Qingjie Fazhan jizhi shichang: Zhengce huanjing yu fazhan xiankuang* (China's Clean Development Mechanism market prior to 2012: Policy environment and development status). Beijing: Chinese Ministry of Environmental Protection.

China Securities Regulatory Commission. 2011. *China Securities Regulatory Commission Annual Report 2011.* Beijing: China Financial & Economic Publishing House.

Chinese Ministry of Environmental Protection (MEP). 2005. *Huanjing yingxiang pingjia gongchengshi zuye zikezheng ji guanli zhixing banfa* (Methods of accreditation for organizations conducting environmental impact assessments). Beijing: MEP.

Chinese National Development and Reform Commission. 2007. *Guanyu fabu xiangmu shenqing baogao* (Notice on disclosing project applications). Beijing: NDRC.

Chinese National People's Congress. [1986–1998] 2004. *Zhonghua Renmin Gongheguo Tudi Guanli Fa* (Land Administration Law of the People's Republic of China). http://www.gov.cn/banshi/2005-05/26/content_989.htm. Accessed February 17, 2012.

———. 2002a. *Zhonghua Renmin Gongheguo Huanjing Yingxiang Pingjia Fa* (Environmental Impact Assessment Law of the People's Republic of China). http://www.people.com.cn/GB/shehui/212/3572/3574/20021029/853043.html. Accessed February 21, 2012.

———. 2002b. *Zhonghua Renmin Gongheguo Shuifa* (Water Law of the People's Republic of China). http://www.mwr.gov.cn/zwzc/zcfg/fl/200210/t20021001_155904.html. Accessed October 10, 2011.

——. 1996. Zhonghua Renmin Gongheguo Dianli Fa (Electric Power Law of the People's Republic of China). http://www.gov.cn/ztzl/2005-12/30/content_142165.htm. Accessed November 1, 2011.

Chinese State Council. 2013. *Guowuyuan guanyu yinfa nengyuan fazhan Shi'er Wu Guihuade tongzhi* (Notice on energy production in the Twelfth Five-Year Plan). Beijing: State Council, January 23.

——. 2012. *Guowuyuan guanyu shixing zui yange shui ziyuan guanli zhidude yijian* (State Council opinion on implementing the strictest water-resource-management system). Beijing: State Council.

——. [1991] 2006. *Da Zhong Xing Shuili Shuidian Gongcheng Jianshe Zhengdi Buchang He Yimin Anzhi Tiaoli* (Regulations on Land-Acquisition Compensation and Resettlement of Migrants for Construction of Large- and Medium-Scale Water-Conservancy and Hydropower Projects). July 7. http://www.mwr.gov.cn/zwzc/zcfg/xzfghfgxwj/200607/t20060707_155925.html.

Clark, Marco. 2009. "Climbing the Mountain Within: Understanding Development Impacts and Understanding Change in Southwest China." M.A. thesis, Oregon State University, Corvallis.

Clarke, Robin and Jannet King. 2004. *The World Atlas: A Unique Visual Analysis of the World's Most Critical Resource.* New York: New Press.

Coggins, Christopher. 2003. *The Tiger and the Pangolin: Nature, Culture, and Conservation in China.* Honolulu: University of Hawai'i Press.

Cooke, Bill and Uma Kothari. 2001. "The Case for Participation as Tyranny." In Bill Cooke and Uma Kothari, eds., *Participation: The New Tyranny?* 1–15. London: Zed Books.

Deng, Xiaoping. [1950] 2006. *Xinan gongzuo wenji* [Collected works on the southwest region]. Chongqing: Chongqing Publishing.

De Soto, Hernando. 2003. *The Mystery of Capital: Why Capitalism Triumphs in the West and Fails Everywhere Else.* New York: Basic Books.

Donahue, John R. and Barbara Rose Johnston. 1998. Conclusion to John R. Donahue and Barbara Rose Johnston, eds., *Water, Culture, and Power: Local Struggles in a Global Context,* 339–345. Washington, D.C.: Island Press.

Dore, John and Xiaogang Yu. 2004. *Yunnan Hydropower Expansion: Update on China's Energy Industry Reforms and the Nu, Lancang, and Jinsha Hydropower Dams.* Working Paper. Chiang Mai, Thailand: Chiang Mai University, United for Social and Environmental Research.

Dore, John, Xiaogang Yu, and K. Y. Li. 2007. "China's Energy Reforms and Hydropower Expansion in Yunnan." In Louis Lebel, John Dore, Rajesh Daniel, and Yang Saing Koma, eds., *Democratizing Water Governance in the Mekong Region,* 55–92. Chiang Mai, Thailand: Mekong Press.

Dowie, Mark. 2009. *Conservation Refugees: The Hundred-Year Conflict Between Global Conservation and Native Peoples.* Cambridge, Mass.: MIT Press.

Dudley, Nigel, ed. 2008. *Guidelines for Applying Protected Area Management Categories.* Gland, Switzerland: International Union for Conservation of Nature.

The Economist. 2012. "The Mekong River: Lies, Dams, and Statistics." July 26. http://www.economist.com/blogs/banyan/2012/07/mekong-river.

Economy, Elizabeth C. 2007. "The Great Leap Backward? The Costs of China's Environmental Crisis." *Foreign Affairs* 86(5): 38–59.

———. 2004. *The River Runs Black: The Environmental Challenge to China's Future.* Ithaca, N.Y.: Cornell University Press.

Edelman, Marc. 2005. "Bringing the Moral Economy Back In . . . to the Study of 21st Century Transnational Social Movements." *American Anthropologist* 107(3): 331–345.

Environmental News Service. 2013. "Keystone XL Pipeline Gets Upbeat Analysis from State Department." Washington, D.C., March 4. http://ens-newswire.com/2013/03/04/keystone-xl-pipeline-gets-upbeat-analysis-from-state-department/.

Erard, Michael. 2009. "How Many Languages? Linguists Discover New Tongues in China." *Science* 324:332–333.

Erni, Christian. 2008. "Country Profile: China." In Christian Erni, ed., *The Concept of Indigenous Peoples in Asia: A Resource Book,* 357–364. Copenhagen: International Working Group for Indigenous Affairs.

Ethnologue. 2009. "Lisu: A Language of China." http://www.ethnologue.com. Accessed July 1, 2009.

Fei, Xiaotong and Zhiyi Zhang. 1945. *Earthbound China: A Study of Rural Economy in Yunnan.* Chicago: University of Chicago Press.

Feng, Yan, Daming He, and Haosheng Bao. 2004. "Analysis on Equitable and Reasonable Allocation Models of Water Resources in the Lancang–Mekong River Basin." *Water International* 29(1): 114–118.

Feng, Yan and Darrin Magee. 2009. "Hyolitical Vulnerability and Resilience in International River Basins in China." In *Hydropolitical Vulnerability and Resilience Along International Waters: Europe, North America, and Asia,* 89–110. Nairobi, Kenya: United Nations Environment Program.

Ferguson, James. 2006. *Global Shadows: Africa in the Neoliberal World Order.* Durham, N.C.: Duke University Press.

———. 1994. *The Anti-politics Machine: "Development," Depoliticization, and Bureaucratic Power in Lesotho.* Minneapolis: University of Minnesota Press.

Field, John. 2003. *Social Capital.* New York: Routledge.

Foster-Moore, Eric. 2011. "Topography and Dams in China." Master's thesis, Oregon State University, Corvallis.

Frey, G. W. and D. M. Linke. 2002. "Hydropower as a Renewable and Sustainable Energy Source Meeting Global Energy Challenges in a Reasonable Way." *Energy Policy* 30(14): 1261–1265.

Galipeau, Brendan A. 2012. "Socio-ecological Vulnerability in a Tibetan Village on the Lancang River, China." Master's thesis, Oregon State University, Corvallis.

Galipeau, Brendan, Mark Ingman, and Bryan Tilt. 2013. "Dam-Induced Displacement and Agricultural Livelihoods in China's Mekong Basin." *Human Ecology* 41(3): 437–446.

Gleick, Peter H. 2009. "China and Water." In Peter H. Gleick, ed., *The World's Water: The Biennial Report on Freshwater Resources, 2008–2009*, 79–100. Washington, D.C.: Island Press.

——. 2003. "Global Freshwater Resources: Soft-Path Solutions for the 21st Century." *Science* 302:1524–1528.

Glover, Denise M., Stevan Harrell, Charles F. McKhann, and Margaret B. Swain, eds. 2011. *Explorers and Scientists in China's Borderlands, 1880–1950*. Seattle: University of Washington Press.

Goodman, David S. J. 2004. "The Campaign to 'Open up the West': National, Provincial-Level, and Local Perspectives." *China Quarterly* 158:317–334.

Goodman, Jim. 2012. *Grand Canyon of the East*. Kunming, China: Yunnan People's Publishing House.

Gorenflo, L. J., Suzanne Romaine, Russell A. Mittermeier, and Kristen Walker-Painemilla. 2012. "Co-occurrence of Linguistic and Biological Diversity in Biodiversity Hotspots and High Biodiversity Wilderness Areas." *Proceedings of the National Academy of Sciences* 109(19): 1–6.

Griffith, David. 2009. "The Moral Economy of Tobacco." *American Anthropologist* 111(4): 432–442.

Gros, Stéphane. 2011. "Economic Marginalization and Social Identity Among the Drung People of Northwest Yunnan." In Jean Michaud and Tim Forsyth, eds., *Moving Mountains: Highland Livelihood and Ethnicity in China, Vietnam, and Laos*, 28–49. Vancouver: University of British Columbia Press.

Grumbine, Edward R. 2010. *Where the Dragon Meets the Angry River: Nature and Power in the People's Republic of China*. Washington, D.C.: Island Press.

Grumbine, Edward R., John Dore, and Jianchu Xu. 2012. "Mekong Hydropower: Drivers of Change and Governance Challenges." *Frontiers in Ecology and the Environment* 10(12): 91–98.

Guo, Huijun, Christine Padoch, Kevin Coffey, Aiguo Chen, and Yongneng Fu. 2002. "Economic Development, Land Use, and Biodiversity Change in the Tropical Mountains of Xishuangbanna, Yunnan, Southwest China." *Environmental Science and Policy* 5(6): 471–479.

Guo, Jiaji. 2008. *Fazhande fansi: Lancangjiang Liuyu shaoshu minzu bianqian de renleixue yanjiu* (Rethinking development: Anthropological studies on the development of ethnic groups in the Lancang–Mekong River basin). Kunming, China: Yunnan Renmin Chubanshe.

Gupta, Akhil and James Ferguson. 1997. "Discipline and Practice: 'The Field' as Site, Method, and Location in Anthropology." In Akhil Gupta and James Ferguson, eds., *Anthropological Locations: Boundaries and Grounds of a Field Science*, 1–33. Berkeley: University of California Press.

Harden, Blaine. 1996. *A River Lost: The Life and Death of the Columbia*. New York: Norton.

Harper, Janice. 2002. *Endangered Species: Health, Illness, and Death Among Madagascar's People of the Forest*. Durham, N.C.: Carolina Academic Press.

Harrell, Stevan. 2011. "Introduction: Explorers, Scientists, and Imperial Knowledge Production in Early Twentieth-Century China." In Denise M. Glover, Stevan Harrell, Charles F. McKhann, and Margaret B. Swain, eds., *Explorers and Scientists in China's Borderlands, 1880–1950*, 3–25. Seattle: University of Washington Press.

——. 1995. "Introduction: Civilizing Projects and the Reaction to Them." In Stevan Harrell, ed., *Cultural Encounters on China's Ethnic Frontiers*, 3–36. Seattle: University of Washington Press.

Harvey, David. 2005. *A Brief History of Neoliberalism*. Oxford: Oxford University Press.

Hathaway, Michael. 2013. *Environmental Winds: Making the Global in Southwest China*. Berkeley: University of California Press.

——. 2010. "The Emergence of Indigeneity: Public Intellectuals and an Indigenous Space in Southwest China." *Cultural Anthropology* 25(2): 301–333.

He, Daming and Lihui Chen. 2002. "The Impact of Hydropower Cascade Development in the Lancang–Mekong Basin, Yunnan." *Mekong Update and Dialogue* 5(3): 2–4.

He, Daming, Jinming Hu, and Yan Feng. 2007. *Zhongguo xinan guoji heliu shui ziyuan liyong yu shengtai baohu* (Utilization of water resources and environmental conservation in the international rivers, southwest China). Beijing: Science Press.

He, Daming, Xiaogang Yu, Lihui Chen, Jiaji Guo, Shu Gan, and Qin Li. 2002. *Master Report of the Study on Manwan Dam Related to the Social, Economic, and Environmental Impacts on the Lancang River*. Hong Kong: Oxfam.

He, Yaohua H., ed. 2008. *Nujiang, Lancangjiang, Jinshajiang: Shuineng ziyuan kaifa yu huanjing baohu yanjiu* (Nujiang, Lancangjiang, and Jinshajiang: Research on the exploitation of hydropower resources and environmental protection). Kunming: China Southwestern Nationalities Research Association.

Hennig, Thomas, Wenling Wang, Yan Feng, Xiaokun Ou, and Daming He. 2013. "Review of Yunnan's Hydropower Development: Comparing Large Hydropower Projects Regarding Their Environmental Implications and Socio-economic Consequences." *Renewable and Sustainable Energy Reviews* 27:585–595.

Hillman, Ben. 2003. "Paradise Under Construction: Minorities, Myths, and Modernity in Northwest Yunnan." *Asain Ethnicity* 4(2): 175–188.

Hruby, Denise. 2013. "Controversial Dam Approved on Cambodia–Laos Border." *Cambodia Daily* (Phnom Penh), October 4.

Hu, Angang. 2003. "Yige Zhongguo, sige shijie: Zhongguo diqu zazhan chayu de bupingdengxing" (One China, four worlds: The inequality of uneven regional development in China). In Angang Hu, Shaoguang Wang, and Jianming Zhou, eds., *Di'erci zhuanxing: Guojia zhidu jianshe* (The second transformation: National institution building), 1–20. Beijing: Tsinghua University Press.

Hu, Angang and Lianhe Hu. 2011. "Di'er dai minzu zhengce: Cujin minzu jiaorong yiti he fanrong yiti" (Second-generation nationalities policy: Promoting ethnic integration and prosperity). *Sociology of Ethnicity* 100:1–14.

Huaneng Corporation. 2010a. *China Huaneng* (Zhongguo Huaneng). Vol. 141. Beijing: Huaneng Corporation.

———. 2010b. *Zhongguo Huaneng* (China Huaneng). Vol. 140. Beijing: Huaneng Corporation.

Imhof, Aviva and Guy R. Lanza. 2010. "Greenwashing Hydropower." *World Watch Magazine* 23(1): 8–17.

Intergovernmental Panel on Climate Change. 2007. *Fourth Assessment Report of the Intergovernmental Panel on Climate Change*. Geneva: Intergovernmental Panel on Climate Change.

International Commission on Large Dams (ICOLD). 1998. *Register of Large Dams*. Paris: ICOLD.

International Law Association. 1966. *The Helsinki Rules on the Uses of the Waters of International Rivers*. Helsinki: International Law Association.

International Rivers. 2013a. "China Moves to Dam the Nu, Ignoring Seismic, Ecological, and Social Risks." Press release, January 25.

———. 2013b. *Spreadsheet of Major Dams in China*. Berkeley, CA: International Rivers.

———. 2012. *The New Great Walls: A Guide to China's Overseas Dam Industry*. 2d ed. Berkeley, Calif.: International Rivers.

International Rivers Network. 2009. "Salween Dams." http://internationalrivers.org /en/southeast-asia/burma/salween-dams. Accessed June 5, 2009.

Interorganizational Committee on Guidelines and Principles for Social Impact Assessment. 2003. "Principles and Guidelines for Social Impact Assessment in the USA." *Impact Assessment and Project Appraisal* 21(3): 231–250.

———. 1994. "Guidelines and Principles for Social Impact Assessment." *Impact Assessment* 12(2): 107–152.

Jia, Xinting and Roman Tomasic. 2009. *Corporate Governance and Resource Security in China*. New York: Routledge.

Johnston, Barbara Rose. 2012. "Water, Culture, Power: Hydrodevelopment Dynamics." In Barbara Rose Johnston, Lisa Hiwasaki, Irene J. Klaver, Ameyali Ramos Castillo, and Veronica Strang, eds., *Water, Cultural Diversity, and Global Environmental Change: Emerging Trends, Sustainable Futures?* 295–318. New York: UNESCO International Hydrological Program.

Johnston, Barbara Rose and John M. Donahue. 1998. Introduction to John R. Donahue and Barbara Rose Johnston, eds., *Water, Culture, and Power: Local Struggles in a Global Context*, 1–5. Washington, D.C.: Island Press.

Kaiser, Brooks A. 2006. "The National Environmental Policy Act's Influence on USDA Forest Service Decision-Making, 1974–1996." *Journal of Forest Economics* 12:109–130.

Kibler, Kelly. 2011. "Development and Decommissioning of Small Dams: Analysis of Impact and Context." Ph.D. diss., Oregon State University, Corvallis.

Kleinman, Arthur, Yunxiang Yan, Jing Jun, Sing Lee, Everett Zhang, Pan Tianshu, Wu Fei, and Guo Jinhua. 2011. *Deep China: The Moral Life of the Person*. Berkeley: University of California Press.

Klose, Christian D. 2012. "Evidence for Anthropogenic Surface Loading as Trigger Mechanism of the 2008 Wenchuan Earthquake." *Environmental Earth Sciences* 66(5): 1439–1447.

LaFargue, Michael. 2001. "'Nature' as Part of Human Culture in Daoism." In N. J. Girardot, James Miller, and Xiaogan Liu, eds., *Daoism and Ecology: Ways Within a Cosmic Landscape*, 45–60. Cambridge, Mass.: Harvard University Press.

Lai, Honyi H. 2002. "China's Western Development Program: Its Rationale, Implementation, and Prospects." *Modern China* 28(4): 432–466.

Lao Tzu. 1997. *Tao Te Ching*. Translated by Gia-Fu Feng and Jane English. New York: Vintage.

Lazrus, Heather. 2009. "The Governance of Vulnerability: Climate Change and Agency in Tuvalu, South Pacific." In Susan A. Crate and Mark Nuttall, eds., *Anthropology and Climate Change: From Encounters to Actions*, 240–249. Walnut Creek, Calif.: Left Coast Press.

Leong, Kirsten M., John F. Forester, and Daniel J. Decker. 2009. "Moving Public Participation Beyond Compliance: Uncommon Approaches to Finding Common Ground." *George Wright Forum* 26(3): 23–39.

Lewis, Joanna. 2013. *Green Innovation in China: China's Wind Power Industry and the Global Transition to a Low-Carbon Economy*. New York: Columbia University Press.

Li, Jing. 2013. "Western-Funded Green Groups 'Stir Up Trouble' in China." *South China Morning Post* (Hong Kong), August 23. http://www.scmp.com/news/china/article/1298716/western-funded-green-groups-stir-trouble-china.

Li, Lianjiang. 2004. "Political Trust in Rural China." *Modern China* 30(2): 228–258.

Li, Qiang and Lingling Shi. 2011. "Shehui yingxiang pingjia ji qi zai wo guo de yingyong" (Social impact assessment and its application in China). *Academics* 156(5): 19–27.

Li, Xinran. 2008. "Nujiang shuidian kaifa dui zengjia dangdi nongmin shourude liyi" (The benefits of hydropower development on the Nu River for increasing local farmers' incomes). In Yaohua H. He, ed., *Nujiang, Lancangjiang, Jinshajiang: Shuineng ziyuan kaifa yu huanjing baohu yanjiu* (Nujiang, Lancangjiang, and Jinshajiang: Research on the exploitation of hydropower resources and environmental protection), 264–279. Kunming: China Southwestern Nationalities Research Association.

Litzinger, Ralph. 2007. "In Search of the Grassroots: Hydroelectric Politics in Northwest Yunnan." In Elizabeth J. Perry and Merle Goldman, eds., *Grassroots Political Reform in Contemporary China*, 282–299. Cambridge, Mass.: Harvard University Press.

——. 2004. "The Mobilization of 'Nature': Perspectives from North-west Yunnan." *China Quarterly* 178:488–504.

Liu, Shouying, Michael R. Carter, and Yang Yao. 1998. "Dimensions and Diversity of Property Rights in Rural China: Dilemmas on the Road to Further Reform." *World Development* 26(10): 1789–1806.

Liu, Zhenya. 2012. *Zhongguo dianli yu nengyuan* (Electric power and energy in China). Beijing: China Electric Power Press.

Lowe, Celia. 2006. *Wild Profusion: Biodiversity Conservation in an Indonesian Archipelago*. Princeton, N.J.: Princeton University Press.

Ma, Jianbo. 2013. *The Land Development Game in China*. New York: Lexington Books.

Ma, Jun. 2004. *China's Water Crisis*. Norwalk, Conn.: Eastbridge Press.

Magee, Darrin. 2011. "The Dragon Upstream: China's Role in Lancang–Mekong Development." In Joakim Öjendal, Stina Hansson, and Sofie Hellberg, eds., *Water, Politics, and Development in a Transboundary Watershed: The Case of the Lower Mekong Basin*, 171–190. New York: Springer.

———. 2006. "Powershed Politics: Hydropower and Interprovincial Relations Under Great Western Development." *China Quarterly* 185:23–41.

Magee, Darrin and Kristen McDonald. 2009. "Beyond Three Gorges: Nu River Hydropower and Energy Decision Politics in China." *Asian Geographer* 25(1–2): 39–60.

Managi, Shunsuke and Shinji Kaneko. 2009. *Chinese Economic Development and the Environment*. Cheltenham, U.K.: Edward Elgar.

Marcott, Shaun A., Jeremey D. Shakun, Peter U. Clark, and Alan C. Mix. 2013. "A Reconstruction of Regional and Global Temperature for the Past 11,300 Years." *Science* 339:1198–1201.

Marx, Karl. [1867] 1984. *Capital: A Critique of Political Economy*. Vol. 1. Translated by Samuel Moore and Edward Aveling. 11th ed. New York: International.

McCully, Patrick. 2001. *Silenced Rivers: The Politics and Ecology of Large Dams*. 2d ed. London: Zed Books.

McDonald, Kristen. 2007. "Damming China's Grand Canyon: Pluralization Without Democratization in the Nu River Valley." Ph.D. diss., University of California, Berkeley.

McDonald, Kristen, Peter Bosshard, and Nicole Brewer. 2009. "Exporting Dams: China's Hydropower Industry Goes Global." *Journal of Environmental Management* 90 (Supplement 3): S294–S302.

Meng, Yang. 2012. "Chinese Power, Burmese Politics." *The Third Pole*, April 2.

Mertha, Andrew C. 2008. *China's Water Warriors: Citizen Action and Policy Change*. Ithaca, N.Y.: Cornell University Press.

Michaud, Jean. 2010. "Editorial: Zomia and Beyond." *Journal of Global History* 5:187–214.

Michaud, Jean and Tim Forsyth, eds. 2011. *Moving Mountains: Ethnicity and Livelihoods in Highland China, Vietnam, and Laos*. Vancouver: University of British Columbia Press.

Miller, Lucien, ed. 1994. *South of the Clouds: Tales from Yunnan*. Translated by Guo Xu, Lucien Miller, and Xu Kun. Seattle: University of Washington Press.

Miller, T. R., T. D. Baird, C. M. Littlefield, G. Kofinas, F. Chapin III, and C. L. Redman. 2008. "Epistemological Pluralism: Reorganizing Interdisciplinary Research." *Ecology and Society* 13(2): 46–62.

Mol, Arthur P. J., and Neil T. Carter. 2006. "China's Environmental Governance in Transition." *Environmental Politics* 15(2): 149–170.

Molle, François, Philippus Wester, Phil Hirsch, Jens R. Jensen, Hammond Murray-Rust, Vijay Paranjpye, Sharon Pollard, and Pieter van der Zaag. 2007. "River Basin Development and Management." In David Molden, ed., *Water for Food, Water for Life: A Comprehensive Assessment of Water Management in Agriculture*, 585–625. London: Earthscan.

Moore, Malcolm 2011. "Chinese Lose Patience with Pollution." *Telegraph* (London), December 6. http://www.telegraph.co.uk/earth/earthnews/8938159/Chinese-lose -patience-with-pollution.html.

Mosse, David. 2005. *Cultivating Development: An Ethnography of Aid and Practice.* London: Pluto Press.

——. 2003. *The Rule of Water: Statecraft, Ecology, and Collective Action in South India.* Oxford: Oxford University Press.

Mueggler, Erik. 2001. *The Age of Wild Ghosts: Memory, Violence, and Place in Southwest China.* Berkeley: University of California Press.

Nader, Laura. 1996. "Introduction: Anthropological Inquiry Into Boundaries, Power, and Knowledge." In Laura Nader, ed., *Naked Science: Anthropological Inquiry Into Boundaries, Power, and Knowledge,* 1–25. New York: Routledge.

Narayan, Ranjana. 2013. "India, China Ink Accord on River Information." *Indo-Asian News Service* (Delhi), October 23.

Naughton, Barry J. 2006. *The Chinese Economy: Transitions and Growth.* Cambridge, Mass.: MIT Press.

Ngo, The Vihn. 2012. "Laos PDR Breaks Ground for Xayaburi Dam: A Tragic Day for the Mekong River and Mekong Delta." Viet Ecology Foundation (Portland, Oregon), December 3. http://www.vietecology.org/Article.aspx/Article/94.

Nilsson, Christer, Catherine A. Reidy, Mats Dynesius, and Carmen Revenga. 2005. "Fragmentation and Flow Regulation of the World's Large River Systems." *Science* 308:798–800.

Nolan, Riall W. 2002. *Development Anthropology: Encounters in the Real World.* Boulder, Colo.: Westview Press.

Nolin, Anne W. and Christopher Daly. 2006. "Mapping 'at Risk' Snow in the Pacific Northwest." *Journal of Hydrometeorology* 7:1164–1171.

O'Brien, Kevin J., and Lianjiang Li. 2007. *Rightful Resistance in Rural China.* Cambridge: Cambridge University Press.

Oi, Jean C. 1999. "Two Decades of Rural Reform in China: An Overview and Assessment." *China Quarterly* 159:616–628.

Oliver-Smith, Anthony. 2010. *Development and Dispossession: The Crisis of Forced Displacement and Resettlement.* Santa Fe: School for Advanced Research Press.

Ong, Aiwha. 2006. *Neoliberalism as Exception: Mutations in Citizenship and Sovereignty.* Durham, N.C.: Duke University Press.

Osborne, Milton. [1975] 1999. *River Road to China: The Search for the Source of the Mekong, 1866–73.* New York: Atlantic Monthly Press.

Palaniappan, Meena and Peter H. Gleick. 2009. "Peak Water." In Peter H. Gleick, ed., *The World's Water: The Biennial Report on Freshwater Resources, 2008–2009,* 1–16. Washington, D.C.: Island Press.

Pan, Yue. 2006. Interview. *China Dialogue,* May 12.

People's Daily (Beijing). 2012. "Wo Da Shuidian Jutou guafen Jinsha Jiang: Bu dao bai gongli you yi zuo shuiku" (Five Hydropower Giants divide up the Jinsha River: A reservoir every hundred kilometers). May 4.

Pieke, Frank. 2009. *The Good Communist: Elite Training and State Building in Today's China.* Cambridge: Cambridge University Press.

Poff, N. Leroy and David D. Hart. 2002. "How Dams Vary and Why It Matters for the Emerging Science of Dam Removal." *Bioscience* 52(8): 59–68.

Polanyi, Michael. [1944] 1957. *The Great Transformation.* Boston: Beacon Press.

——. 1962. "The Republic of Science: Its Political and Economic Theory." *Minerva* 1:54–74.

Post (Zambia). 2010. "Kafue Gorge Lower Hydropower Station." August 24. http://www.postzambia.com/post-read_article.php?articleId=13002.

Postel, Sandra and Brian D. Richter. 2003. *Rivers for Life: Managing Water for People and Nature.* Washington, D.C.: Island Press.

Pottinger, Lori. 2012. "Why Big Dams Don't Work." *The Third Pole,* July 11. http://www.chinadialogue.net/article/show/single/en/5037-Why-big-dams-don-t-work.

Rapley, John. 2007. *Understanding Development: Theory and Practice in the Third World.* 3rd ed. London: Lynne Rienner.

Reuters Africa. 2012. "World Bank Approves Loan for Cameroon Dam Project." March 28. http://af.reuters.com/article/investingNews/idAFJOE82R00I20120328.

Ribot, Jesse C. and Nancy Lee Peluso. 2003. "A Theory of Access." *Rural Sociology* 68(2): 153–181.

Richter, Brian D., Sandra Postel, Carmen Revenga, Thayer Scudder, Bernhard Lehner, and Allegra Churchill. 2010. "Lost in Development's Shadow: The Downstream Human Consequences of Dams." *Water Alternatives* 3(2): 14–42.

Rock, Joseph F. 1947. *The Ancient Nakhi Kingdom of Southwest China.* Cambridge, Mass.: Harvard University Press.

Rossabi, Morris. 2004. Introduction to Morris Rossabi, ed., *Governing China's Multiethnic Frontiers,* 3–18. Seattle: University of Washington Press.

Rozelle, Scott, Loren Brandt, Li Guo, and Jikun Huang. 2005. "Land Tenure in China: Facts, Fictions, and Issues." In Peter Ho, ed., *Developmental Dilemmas: Land Reform and Institutional Change in China,* 121–150. New York: Routledge.

Sadler, Barry, Frank Vanclay, and Iara Verocai. 2000. *Environmental and Social Impact Assessment for Large Dams: World Commission on Dams Thematic Review.* Vol. 2. Cape Town: World Commission on Dams.

Santasombat, Yos. 2011. *The River of Life: Changing Ecosystems of the Mekong Region.* Chiang Mai, Thailand: Mekong Press.

Save the Mekong Coalition. 2010. "Call for the Complete Release of Data on the 2010 Mekong Region Drought." Letter to the Mekong River Commission, June 15.

Schnitter, Nicholas J. 1994. *A History of Dams: The Useful Pyramids.* Leiden: Balkema.

Scott, James C. 2009. *The Art of Not Being Governed: An Anarchist History of Upland Southeast Asia.* New Haven, Conn.: Yale University Press.

——. 2005. "Afterword to 'Moral Economies, State Spaces, and Categorical Violence.'" *American Anthropologist* 107(3): 395–402.

——. 1998. *Seeing Like a State: How Certain Schemes to Improve the Human Condition Have Failed.* New Haven, Conn.: Yale University Press.

———. 1976. *The Moral Economy of the Peasant.* New Haven, Conn.: Yale University Press.

Scudder, Thayer. 2005. *The Future of Large Dams: Dealing with Social, Environmental, Institutional, and Political Costs.* London: Earthscan.

Selden, Mark. 1998. "House, Cooperative, and State in the Remaking of China's Countryside." In Eduard B. Vermeer, Frank N. Pieke, and Woel Lien Chong, eds., *Cooperative and Collective in China's Rural Development: Between State and Private Interests,* 17–45. Armonk, N.Y.: M. E. Sharpe.

Shapiro, Judith. 2001. *Mao's War Against Nature: Politics and the Environment in Revolutionary China.* Cambridge: Cambridge University Press.

Shi, Jiangtao. 2009. "Wen Calls Halt to Yunnan Dam Plan." *South China Morning Post* (Beijing), May 21.

Sinohydro. 2011. "Key Revenue Figures." http://eng.sinohydro.com/index.php?m=content&c=index&a=lists&catid=18. Accessed April 30, 2013.

Sivaramakrishnan, K. 2005. "Introduction to 'Moral Economies, State Spaces, and Categorical Violence.'" *American Anthropologist* 107(3): 321–330.

———1999. *Modern Forests: Statemaking and Environmental Change in Colonial Eastern India.* Stanford: Stanford University Press.

Smil, Vaclav. 2004. *China's Past, China's Future: Energy, Food, Environment.* New York: Routledge.

Sommer, Deborah. 2012. "Ritualized Landscapes? Conceptualizations of the Earth and the Ritual Management of Natural Resources in Classical Chinese Texts." In *Conference Proceedings: Religious Diversity and Ecological Sustainability in China,* 351–366. Beijing: Minzu University of China.

Stender, Neal, Wang Dong, and Jing Zhou. 2002. "The New EIA Law and Environmental Protection in China." *China Law and Practice* 16(10): 1–5

Stern, Rachel E. 2011. "From Dispute to Decision: Suing Polluters in China." *China Quarterly* 206:294–312.

Stone, Elizabeth C. and Paul E. Zimansky. 2004. *The Anatomy of a Mesopotamian City: Survey and Soundings at Mashkan-shapir.* Winona Lake, Ind.: Eisenbraun's.

Swinnen, Johan F. M. and Scott Rozelle. 2006. *From Marx and Mao to the Market: The Economics and Politics of Agricultural Transition.* Oxford: Oxford University Press.

Taiban, Sasala. 2004. *Xunzhao shiluode jianshi: Buluo zhuyide shiye je xingdong* (Seeking the lost arrow: The vision and action of tribalism). Taipei: National Development Foundation for Culture and Education.

Tang, Shuiyan and Xueyong Zhan. 2008. "Civic Environmental NGOs, Civil Society, and Democratization in China." *Journal of Development Studies* 44(3): 425–448.

Tang, Wentao. 2007. "Zhongguo Huanjing Yingxiang Pingjia Fa de guanli chengxu fenxi" (An analysis of the management procedures in China's Environmental Impact Assessment Law). *Journal of Changzhou Institute of Engineering Technology* 3(53): 50–52.

Thomas, Sebastian, Paul Dargusch, and Andrew Griffiths. 2011. "The Drivers and Outcomes of the Clean Development Mechanism in China." *Environmental Policy and Governance* 21:223–239.

Thompson, E. P. 1991. *The Making of the English Working Class*. Toronto: Penguin Books.

———. 1971. "The Moral Economy of the English Crowd in the 18th Century." *Past and Present* 50:76–136.

Tilt, Bryan. 2011. "View from the 21st Century: Civil Society and Environment in China." In Barbara Rose Johnston, ed., *Life and Death Matters: Human Rights, Environment, and Social Justice*, 2nd ed., 62–76. Walnut Creek, Calif.: Left Coast Press.

———. 2010. *The Struggle for Sustainability in Rural China: Environmental Values and Civil Society*. New York: Columbia University Press.

Tilt, Bryan, Yvonne A. Braun, and Daming He. 2009. "Social Impact Assessment for Large Dam Projects: A Comparison of International Projects and Implications for Best Practice." *Journal of Environmental Management* 90 (Supplement 3): 249–257.

Tilt, Bryan and John A. Young. 2007. "Introductory Essay: Development in Contemporary China." *Urban Anthropology and Studies of Cultural Systems and World Economic Development* 36(1–2): 1–8.

Tsing, Anna Lowenhaupt. 2005. *Friction: An Ethnography of Global Connection*. Princeton, N.J.: Princeton University Press.

Tu, Weiming. 1998. "The Continuity of Being: Chinese Visions of Nature." In Mary Evelyn Tucker and John Berthrong, eds., *Confucianism and Ecology: The Interrelation of Heaven, Earth, and Humans*, 105–122. Cambridge, Mass.: Harvard University Press.

Tucker, Mary Evelyn and Duncan Ryuken Williams, eds. 1998. *Buddhism and Ecology: The Interconnection of Dharma and Deeds*. Cambridge, Mass.: Harvard University Press.

Tullos, Desiree, Phil H. Brown, Kelly Kibler, Darrin Magee, Bryan Tilt, and Aaron T. Wolf. 2010. "Perspectives on the Salience and Magnitude of Dam Impacts for Hydro-development Scenarios in China." *Water Alternatives* 3(2): 71–90.

Tullos, Desiree D., Eric Foster-Moore, Darrin Magee, Bryan Tilt, Aaron T. Wolf, Edwin Schmitt, Francis Gassert, and Kelly Kibler. 2013. "Biophysical, Socioeconomic, and Geopolitical Vulnerabilities to Hydropower Development on the Nu River, China." *Ecology and Society* 18(3): 16–33.

Turner, Jennifer L. 2005. "River Basin Governance in China." *China Environment Forum* 7:106–110.

United Nations Development Program (UNDP). 2008. *Human Development Report: China*. Beijing: UNDP.

United Nations Educational, Scientific and Cultural Organization (UNESCO). 2003. *World Heritage Nomination—IUCN Technical Evaluation: Three Parallel Rivers of Yunnan Protected Areas*. Paris: UNESCO.

United Nations Environment Program (UNEP). 2009. *Three Parallel Rivers of Yunnan Protected Areas*. Paris: UNEP.

——. 2008. *UN Water Status Report on IWRM and Water Efficiency Plans*. Paris: UNEP.

——. 1992. *Rio Declaration on Environment and Development*. Rio de Janeiro: UNEP.

United Nations General Assembly. 1997. *Convention on the Law of the Non-navigational Uses of International Watercourses*. New York: United Nations.

——. 1986. *Declaration on the Right to Development*. 79th Plenary Meeting, December 4. New York: United Nations.

——. 1947. *Universal Declaration of Human Rights*. New York: United Nations.

United Nations Permanent Forum on Indigenous Issues. 2007. *Declaration on the Rights of Indigenous Peoples*. New York: United Nations. http://www.un.org/esa /socdev/unpfii/. Accessed June 24, 2009.

United Nations World Heritage Committee. 2004. *Decisions Adopted at the 28th Session of the World Heritage Committee*. WHC-04/28 COM. Suzhou, China: United Nations World Heritage Committee.

U.S. National Research Council, Committee on Himalayan Glaciers, Hydrology, Climate Change, and Implications for Water Security. 2012. *Himalayan Glaciers: Climate Change, Water Resources, and Water Security*. Washington, D.C.: National Academies Press.

U.S. National Science Foundation, Technical Advisory Committee for Environmental Research and Education. 2009. *Transitions and Tipping Points in Complex Environmental Systems*. Washington, D.C.: U.S. National Science Foundation.

Vanclay, Frank. 2003. "International Principles for Social Impact Assessment." *Impact Assessment and Project Appraisal* 21(1): 5–11.

——. 2002a. "Conceptualizing Social Impacts." *Environmental Impact Assessment Review* 22(3): 183–211.

——. 2002b. "Social Impact Assessment." *Encyclopedia of Global Environmental Change* 4:387–393.

Van Deth, Jan W. 2003. "Measuring Social Capital: Orthodoxies and Continuing Controversies." *International Journal of Social Science Research Methodology* 6(1): 79–92.

Van Schendel, Willem. 2002. "Geographies of Knowing, Geographies of Ignorance: Jumping Scale in Southeast Asia," *Environment and Planning D: Society and Space* 20(6): 647–668.

Vidal, John and David Adam. 2007. "China Overtakes U.S. as World's Biggest CO_2 Emitter." *Guardian* (London), June 19.

Wang, Alex. 2007. *One Billion Enforcers*. Washington, D.C.: Environmental Law Institute.

Wang, Jianjun, Genya Yu, and Shuifeng Li. 2008. "Meiguo daba jingji fazhan dui Nujiang shuidian kaifade yingxiang: zai Meiguode daba jianchu" (The implications of the development of a dam economy in the United States for Nu River hydropower development: Dam removals in the United States). In Yaohua H. He, ed., *Nujiang, Lancangjiang, Jinshajiang: Shuineng ziyuan kaifa yu huanjing baohu yanjiu* (Nujiang, Lancangjiang, and Jinshajiang: Research on the exploitation of hydropower resources and environmental protection), 148–186. Kunming: China Southwestern Nationalities Research Association.

Wang, Jianming and John A. Young. 2006. "Applied Anthropology in China." *National Association for Practicing Anthropology Bulletin* 25:70–81.

Wang, Pu, Steven A. Wolf, James P. Lassoie, and Shikui Dong. 2013. "Compensation Policy for Displacement Caused by Dam Construction in China: An Institutional Analysis." *Geoforum* 48:1–9.

Wang, Shaoguang and Angang Hu. 1999. *The Political Economy of Uneven Development: The Case of China.* Armonk, N.Y.: M. E. Sharpe.

Ward, Francis Kingdon. 1923. *The Mystery Rivers of Tibet.* London: Seeley, Service.

Watson, Nigel. 2004. "Integrated River Basin Management: A Case for Collaboration." *International Journal of River Basin Management* 2(4): 243–257.

Watts, Jonathan. 2011. "China Warns of 'Urgent Problems' Facing Three Gorges Dam." *Guardian* (Beijing), May 20.

Webber, Michael and Brooke McDonald. 2004. "Involuntary Resettlement, Production, and Income: Evidence from Xiaolangdi, PRC." *World Development* 32(4): 673–690.

Weller, Robert P. 2006. *Discovering Nature: Globalization and Environmental Culture in China and Taiwan.* Cambridge: Cambridge University Press.

——, ed. 2005. *Civil Life, Globalization, and Political Change in Asia: Organizing Between Family and State.* London: Routledge.

West, Paige. 2006. *Conservation Is Our Government Now: The Politics of Ecology in Papua New Guinea.* Durham, N.C.: Duke University Press.

West, Paige and Daniel Brockington. 2006. "An Anthropological Perspective on Some Unexpected Consequences of Protected Areas." *Conservation Biology* 20(3): 609–616.

Whiting, Susan. 2011. "Values in Land: Fiscal Pressures, Land Disputes, and Justice Claims in Rural and Peri-urban China." *Urban Studies* 48(3): 569–587.

Wilkes, Andreas J. 2005. "Ethnic Minorities, Environment, and Development in Yunnan: The Institutional Contexts of Biocultural Knowledge Production in Southwest China." Ph.D. diss., University of Kent.

Wilson, Edward O. 2007. *The Creation: An Appeal to Save Life on Earth.* New York: Norton.

Wisner, Ben, Piers Blaikie, Terry Cannon, and Ian Davis. 2004. *At Risk: Natural Hazards, People's Vulnerability, and Disasters.* London: Routledge.

Wolf, Aaron T., ed. 2009. *Hydropolitical Vulnerability and Resilience Along International Waters: Asia.* Nairobi: United Nations Environment Program.

Wolf, Aaron T., Jeffrey A. Natharius, and Jeffrey J. Danielson. 1999. "International River Basins of the World." *International Journal of Water Resources Development* 15(4): 387–427.

Wolf, Aaron T., Shira B. Yoffe, and Mark Giordano. 2003. "International Waters: Identifying Basins at Risk." *Water Policy* 5(1): 29–60.

Wolf, Eric. 1990. "Distinguished Lecture: Facing Power—Old Insights, New Questions." *American Anthropologist* 92:586–596.

World Bank. 2005. *Community-Driven Development.* New York: World Bank. http://go.worldbank.org/24K8IHVVS0. Accessed May 14, 2009.

World Commission on Dams (WCD). 2000. *Dams and Development: A New Framework for Decision-Making (Report Summary)*. London: Earthscan.

World Nuclear Association. 2013. *Comparison of Life Cycle Greenhouse Gas Emissions of Various Electricity Generation Sources*. London: World Nuclear Association.

Worldwatch Institute. 2001. *Liberating the Rivers*. Washington, D.C.: Worldwatch Institute.

World Wide Fund for Nature. 2004. *Rivers at Risk: Dams and the Future of Freshwater Ecosystems*. Gland, Switzerland: World Wide Fund International.

Worster, Donald. 2011. *The Flow of Empire: Comparing Water Control in China and the United States*. Munich: Rachel Carson Center Perspectives.

——. 1985. *Rivers of Empire: Water, Aridity, and the Growth of the American West*. New York: Pantheon Books.

Wutich, Amber. 2011. "The Moral Economy Reexamined: Reciprocity, Water Insecurity, and Urban Survival in Cochabamba, Bolivia." *Journal of Anthropological Research* 67(1): 5–26.

Xu, Jianchu and David R. Melick. 2007. "Rethinking the Effectiveness of Public Protected Areas in Southwestern China." *Conservation Biology* 21(2): 318–328.

Xu, Jianchu and Andreas Wilkes. 2004. "Biodiversity Impact Analysis in Northwest Yunnan, Southwest China." *Biodiversity and Conservation* 13:959–983.

Yang, Guobin. 2005. "Environmental NGOs and Institutional Dynamics in China." *China Quarterly* 181:46–66.

Yang, Mayfair. 1994. *Gifts, Favors, and Banquets: The Art of Social Relationships in China*. Ithaca, N.Y.: Cornell University Press.

Yao, R. F. 2004. *Shui Li Bu: Wo guo shuiku yimin zongshu 1,500 duo wan* (Ministry of Water Resources: Our nation's dam-induced migrants total at least 15 million). Beijing: Xinhua News.

Yeh, Emily T. 2014. "Reverse Environmentalism: Contemporary Articulations of Tibetan Buddhism, Culture, and Environmental Protection." In *Religious Diversity and Ecological Sustainability in China*, ed. Peter Van der Veer, James Miller, and Dan Smyer Yu, 194–218. New York: Routledge. Beijing: Minzu University of China.

Yeh, Emily T. and Joanna I. Lewis. 2004. "State Power and the Logic of Reform in China's Energy Sector." *Pacific Affairs* 77(3): 437–465.

Yin, Lun. 2012. "Water Knowledge, Use, and Governance: Tibetan Participatory Development Along the Mekong (Lancang) River, in Yunnan, China." In Barbara R. Johnston, Lisa Hiwasaki, Irene J. Klaver, Ameyali Ramos Castillo, and Veronica Strang, eds., *Water, Cultural Diversity, and Global Environmental Change: Emerging Trends, Sustainable Futures?* 185–201. New York: UNESCO International Hydrological Program.

Ying, Xing. 2005. *The Story of Dahe River Migrants' Petition: From "Seeking an Explanation" to "Rationalization"* (Dahe yimin shangfang de gushi: Cong "tao ge shuofa" dao "baiping lishun"). Toronto: Three Gorges Probe.

Yoshinaga, Alvin, Jiangyu He, Paul Weissich, Paul Harris, and Margaret B. Swain. 2011. "Classifying Joseph Rock: Metamorphic, Conglomerate, and Sedimentary." In

Denise M. Glover, Stevan Harrell, Charles F. McKhann, and Margaret B. Swain, eds., *Explorers and Scientists in China's Borderlands, 1880–1950*, 116–148. Seattle: University of Washington Press.

Yuksel, I. 2007. "Development of Hydropower: A Case Study in Developing Countries." *Energy Sources* 2:113–121.

Yunnan Statistical Bureau. 2011. *Yunnan tongji nianjian* (Yunnan statistical yearbook). Kunming: China Statistics Press.

Yunnan Wang. 2009. "Yunnan Fengqing Shanti huapo qu zhankai shuimian soujiu" (Water rescue under way in landslide area of Fengqing County, Yunnan). July 23.

Zhang, Li. 2010. *In Search of Paradise: Middle-Class Living in a Chinese Metropolis.* Ithaca, N.Y.: Cornell University Press.

Zigon, Jarrett. 2008. *Morality: An Anthropological Perspective.* New York: Berg.

INDEX